Ousmane Doucoure

Impact of solid waste on groundwater in Bamako

Ousmane Doucoure

Impact of solid waste on groundwater in Bamako

Analysis of interactions between urban pollution and water resource quality in the Bamako district

ScienciaScripts

Imprint
Any brand names and product names mentioned in this book are subject to trademark, brand or patent protection and are trademarks or registered trademarks of their respective holders. The use of brand names, product names, common names, trade names, product descriptions etc. even without a particular marking in this work is in no way to be construed to mean that such names may be regarded as unrestricted in respect of trademark and brand protection legislation and could thus be used by anyone.

Cover image: www.ingimage.com

This book is a translation from the original published under ISBN 978-620-3-41491-2.

Publisher:
Sciencia Scripts
is a trademark of
Dodo Books Indian Ocean Ltd. and OmniScriptum S.R.L publishing group

120 High Road, East Finchley, London, N2 9ED, United Kingdom
Str. Armeneasca 28/1, office 1, Chisinau MD-2012, Republic of Moldova, Europe
Managing Directors: Ieva Konstantinova, Victoria Ursu
info@omniscriptum.com

Printed at: see last page
ISBN: 978-620-8-59724-5

DEDICATION

To **Allah** Almighty, to Him belongs the merit and recognition in the accomplishment of this work as well as its benefits in my life ;

To the prophet MOHAMED (peace be upon him)

To my father Souleymane,

You have made real sacrifices for our education. Your prayers have finally been answered, and I also ask God to grant you long life and good health.

To my mother Penda DIALLO,

Who has worked for my success, through his love, his support, all the sacrifices he has made and his invaluable advice, for all his assistance and his presence in my life, receive through this work, however modest it may be, the expression of my feelings and my eternal gratitude.

To my sister, the late Kadidia DOUCOURE,

You left us unprepared for your death. I don't have much to say to express my love for you. I was shattered by your passing on August 31, 2015. I know you rest in a more peaceful place than here. Peace to your soul my dearest sister.

To my dear brothers and sisters,

Thank you for your encouragement and your support, I find here the expression of my infinite love;

To all my family,

To all the members of **DOUAOU**, for the moments of sorrow and joy we shared. And **to the late Youssouf DIALLO**, we'll never forget you wherever you've gone, may God welcome you into his paradise.

And to all my friends around the world, I wish you happiness r

ACKNOWLEDGEMENTS

First of all, we thank **Allah** for giving us the good health we need to complete our graduation project.

We would also like to thank our **parents** and **family** for their support and encouragement during our studies.

Many thanks,

To **Professor Adama TOLOFOUDYE,**

for authorizing me to carry out the analyses in his laboratory

To **Dr Yacouba MAIGA**

For having spontaneously agreed to guide me in my work, and for his support, guidance, encouragement, human qualities and advice.

My thanks go to the members of the **jury** for agreeing to judge this work. My thanks go to all our teachers and trainers at the Institut de Pédagogie Universitaire. My thanks and gratitudes also go to all the staff at the FST/USTTB measurement center.

I am also grateful to my Master's colleagues for all the good times we shared together throughout the training period. Finally, I would like to thank all the people who contributed in any way to my training and to the completion of this thesis, and whose names do not appear in this document. May they all find here the expression of my deep esteem and gratitude.

I would also like to thank all the staff at the Institut de pédagogie universitaire (IPU)

My thanks also go to all my brothers, friends and acquaintances for their support.

Table of contents

SUMMARY

In Bamako, the management of household solid waste is intensifying with population growth and changes in production and consumption patterns. The generation of this waste is inherent to human activities, whether domestic, agricultural, industrial or commercial, and has a strong impact on the environment, for example: waste is carried away by the wind, attracts harmful insects and pollutants such as leachates which can penetrate and contaminate groundwater and watercourses.

The present study is based on the impact of waste on groundwater quality in the Bamako district. Several methods and tools were used in the course of this work, including: an inventory of the various transit dumps in the six communes of the Bamako district, waste characterization, physico-chemical soil analyses, physico-chemical and heavy metal analyses of landfill leachates and water from the wells surrounding the dumps were analyzed at the FST/USTTB measuring center located on Badalabougou hill. The results of these analyses showed that solid waste is made up of multiple elements (plastic, glass, fabrics, wood, etc.), and plastic waste categories are omnipresent in Bamako's landfills. The results of physico-chemical analyses of the soils revealed that they are rich in metallic elements and act as filters for metal retention. Analyses of landfill leachates showed that these waters have an alkaline pH, with values ranging from 7.33 to 7.96. The leachates are contaminated with nitrate, chloride and heavy metals, and have a conductivity well above the maximum values for wastewater discharge in Mali. Analyses of the water in the wells surrounding the landfill have confirmed the presence of nitrates and nitrites and a number of metallic elements in the groundwater. This contamination is due to the infiltration of landfill leachates into the groundwater through percolation. The proposed strategies could help raise awareness among people living near landfills, so that they do not consume untreated well water.

Keywords: Waste, Leachate, pollution, impact, Bamako

I. GENERAL INTRODUCTION

The accelerated urbanization of developing countries, including Mali, the low level of education of the population and the weakness of public sanitation structures, make waste management an enormous challenge. Waste is an inseparable part of human life. From time immemorial and in all places, waste production has been inherent to human activities, whether domestic, agricultural, industrial or commercial (Ouattara et al, 2017). However, waste management issues have evolved over time.

In 2005, the United Nations Environment Programme (UNEP) explained that the growth of urban waste represented a problem not only for developed countries, but also for developing countries grappling with the difficulties of waste management and disposal (Chunyan, 2016).

For a long time, waste was mainly organic in nature, reflecting a rural lifestyle: animal and human dejecta, the remains of slaughtered animals, agricultural residues. Initially, waste was dumped in landfill sites close to where it was produced (N'go and Regent, 2004). The issue of waste was limited to the abandonment of organic matter. But as more and more people settled in a given area, the quantity of waste produced gradually became a nuisance, and the balance of the environment was upset (Ouattara et *al*, 2017).

Since 2006, the Government of Mali and its partners have been engaged in the process of drawing up Mali's National Sanitation Policy (Politique Nationale d'Assainissement du Mali - PNA). This National Sanitation Policy document was adopted in January 2008. It complements the National Environmental Protection Policy, drawn up and adopted in 1998. The National Sanitation Policy is based on five (05) major axes, namely (C.A.E.P, 2018):

Φ Liquid waste management strategy ;

Φ The solid waste management strategy ;

Φ Special waste management strategy ;

Φ Stormwater management strategy ;

Φ Hygiene and sanitation education.

Environmental protection, particularly sanitation, is a major political issue in Mali. Article 15 of the Constitution of February 25, 1992 states that "Everyone has the right to a healthy environment. The protection and defense of the environment and the promotion of the quality of life are a duty for all and for the State". In line with this new Constitution, Mali drew up an ambitious National Environmental Protection Policy (PNPE) in 1998, which highlights the need to "improve the living environment of rural and urban populations by equipping urban and rural centers with sanitation infrastructures and combating various forms of pollution, particularly those affecting water resources" (C.A.E.P, 2018)

The deterioration in the quality of water resources is just as great a threat as that posed by the growing water deficit. Groundwater, once of good quality, is now threatened by various sources of point-source pollution. The percolation of waste leachate is a source or vector of pollution of groundwater resources through infiltration, and surface water through rainwater runoff (Labadi, and Saadoudi, 2019).

The aim of this study is to assess the impact of solid waste on the quality of groundwater surrounding transit depots in the Bamako district.

- **Objectives :**
- **General objective:**

Study the impact of waste on the quality of living groundwater near the Bamako transit depots.

- **Specific objectives :**
  - Inventory transit depots in the communes of the Bamako district;
  - characterize waste ;
  - characterize waste leachates ;
  - assess the influence of waste on the quality of drinking water in the vicinity of Bamako's transit depots.

II. BIBLIOGRAPHICAL REVIEW

2.1. Waste in general

2.1.1. Definition of waste :

Today, in everyday language, waste refers to the loss that something experiences in terms of its volume, value or any of its qualities. The various definitions of waste insist on this notion of loss (Chunyan, 2016).

In the Republic of Mali, waste is defined as "any residue of a production, transformation or use process, any substance, materials produced or, more generally, any movable property abandoned or which its holder intends to abandon and which is likely to produce harmful effects on the soil, flora and fauna, to degrade sites or landscapes, to pollute the air or water, to generate noise or odors, and generally to harm human health and the environment". (Law n°2021/032 of May 21, 2021).

2.1.2. Waste classification :

Depending on the type of waste

There are three (03) essential categories: solid, liquid, and gaseous (Zedam and Laraba, 2019). Given the context and purpose of our study, only solid and semi-solid wastes are considered below:

a. Household and similar waste :

All waste from households and similar waste from industrial, commercial and craft activities which, by its nature and composition, can be assimilated to household waste (Zedam and Laraba, 2019).

b. Special waste :

All waste from industrial, agricultural, healthcare, service and other activities which, due to its nature and the composition of the materials it contains, cannot be collected, transported and treated under the same conditions as household and similar waste and inert waste (Zedam and Laraba, 2019).

c. Hazardous special waste (HHW)

Any special waste which, by its constituents or by the characteristics of the harmful materials it contains, is likely to harm public health and/or the environment (Zedam and Laraba, 2019).

d. Healthcare waste: biomedical waste

All waste from diagnostic, monitoring and preventive or curative treatment activities in the fields of human and veterinary medicine that presents a physical hazard or biological or chemical contamination for humans and/or the environment (Kanté, 2019).

e. Inert waste

Any waste that does not produce a physical or chemical reaction, such as waste from quarrying, mining, demolition, construction or renovation work, and that is not made up of or contaminated by hazardous substances or other nuisance-generating elements (Siham, 2012).

f. Ultimate waste

Ultimate waste is defined as waste that can no longer be treated under the technical and economic conditions required for its recovery or for the reduction of its polluting or hazardous nature (Djemaci, 2012).

g. Bulky waste

All household waste which, because of its bulky nature, cannot be collected under the same conditions as household and similar waste, such as: sofas, armchairs, tables and old furniture (Zedam and Laraba, 2019).

h. Biodegradable waste

This waste is generally of plant or animal origin (green waste or biowaste), and can be decomposed by living organisms over time (Bouazza, 2011).

i. Agricultural waste :

They come from farms, livestock farms or agri-food industries industries (Nawel, 2007).

2.1.3. Characterization of solid waste :

Waste characterization is an operation that provides information on the composition, nature and behavior of the waste produced, so that it can be effectively managed. In other words, the analysis and characterization of waste enables us to identify the criteria to be taken into account in the choice of waste treatment and disposal methods. Characterization takes into account the production environment, the frequency of production and the nature of the waste (Sissoko, 2016).

2.1.4. Waste life :

The lifespan of a waste product corresponds to the time it takes to decompose and disappear into Nature. This time depends on the nature of the waste. In addition to the visual pollution it causes, waste abandoned in the open air represents a real danger to the environment, as it can take an extremely long time to decompose naturally. As it degrades, it can release molecules into the soil, water and air that are hazardous to all living organisms.

Biodegradable organic waste disappears in less than a year. Apple and yam peelings can decompose and disappear into nature within 1 to 5 months.

The decomposition time of a cigarette butt discarded in the environment is around 5 years. Thrown into a stream, it can pollute 500 liters of water. And it takes 10 years for metal to rust and disappear into nature (Quesnel, 2001).

The time it takes for plastics to decompose and disappear into nature ranges from 100 to 1000 years. Plastic decomposition releases microplastics into rivers and oceans, disrupting the trophic chain of aquatic life. This affects the quality of fish products. Glass waste has a degradation time of 4,000 years (Quesnel, 2001).

Table 1: Lifetimes of selected waste products (Ali and Farouk, 2022)

Type of waste	Average service life
▪ Paper handkerchief	▪ 3 months
▪ Newspape	▪ 3 to 12 months
▪ Matches	▪ 6 months
▪ Aluminium can	▪ 200 to 500 years
▪ Plastic bag	▪ 450 years old
▪ Phone card	▪ 1,000 years

2.1.5. Waste nomenclature.

The waste nomenclature is a classification that includes special waste, including hazardous special waste, assimilated household waste and inert waste. It helps the various partners involved in waste disposal to speak a common language.

Waste is classified in a nomenclature that assigns each waste a code allowing a common language at European level, this nomenclature is composed of 20 categories made according to the activity of origin of the waste (Zedam and Laraba, 2019):

- The "A" code number consists of six digits: the first two (sector of activity), the next two (origin and nature of the waste), and the last two (precise characterization of the waste).
- B" the identification of the waste class to which the waste belongs.
- C" identification of the hazardous nature of the waste

The nomenclature must be used by any individual or legal entity involved in waste disposal, i.e.: public authorities, public institutions, local authorities, companies producing, importing, exporting or installing infrastructures for the incineration, physico-chemical or biological treatment of waste.

2.1.6 Impacts of solid waste on health and the environment

Poor waste management can affect the economy in a number of ways, including reduced food production, poor human and animal health, and reduced tourism potential. Incompetent use of resources affects economic efficiency and the ability to produce the food and staple goods required to meet the needs of growing populations. The proliferation of waste, a reflection of our lifestyles, leads to wastage of raw materials and energy, and degradation of the natural environment. We therefore need to manage this waste (Zedam and Laraba, 2019).

The decomposition of solid waste (the organic elements it contains) under the action of water, air and temperature is likely to cause immediate or remote incalculable dangers to the human environment (pollution, bad smells...). The phenomenon is quite serious when waste is poorly managed, as it is composed of fermentable, recyclable, hazardous, inert and plastic materials (Diabate, 2010). Pollution of wells by landfill waste in figure 1.

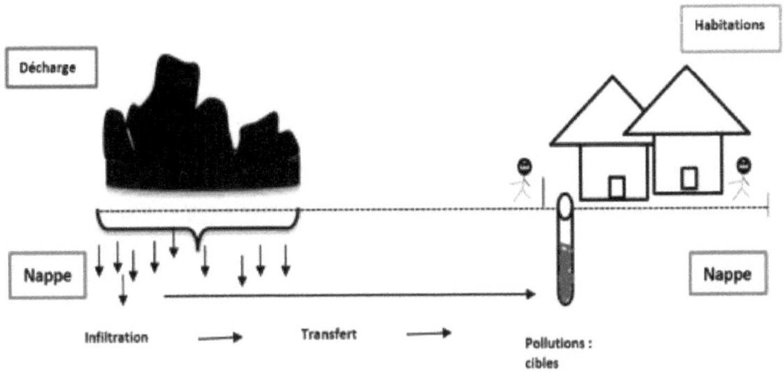

Figure 1: Conceptual diagram of well pollution by landfill waste (Adamou, 2021).

2.1.7 Measuring the quantity of household waste :

Household waste production per inhabitant per day varies widely from one region to another (and also between urban and rural areas). Variations can exist within the same neighborhood, depending on the type of housing: high-standard, medium-standard or low-cost (Quesnel, 2001). The daily production of household waste in the city of Bamako was 2,100 m3/day in 2005, with a collection rate of 60% (BM, 2011). According to the platform of Bamako's partners at Anger (2011), the annual production of solid waste was estimated by the Bamako District Council at 377,775 tonnes in 2011, and the evaluation of existing waste in stock throughout the city of Bamako is 224,054.6 m3.

An assessment of solid waste production in the communes of Bamako for the period 1993 - 2003 was carried out by Tecsult in 1994. The table below gives the average solid waste production of the communes. Hospital and industrial waste are not included (Adamou, 2021).

Table 2: Average household waste production in Bamako from 1993 to 2003 (Adamou, 2021).

	Average production		
	1993	**1998**	**2003**

Municipality	m^3 % m^3 m(3) % m(3)	% m^3 % m(3) % m(3) % m(3)	% m^3	%
Commune I	84 680 15 116 618 16 159 542 15			
Commune II	74 059 14 850 739 11 98 769 10			
Commune III	130,415 24 154,359 21 182,500 18			
Commune IV	94 024 17 13 451 17 172 025 17			
Commune V	81 249 15 126 536 17 190 165 18			
Commune VI	76 176 14 130 086 17 221 117 21			

2.1.8. Determining the Ratio

The ratio (r) is the average quantity of household waste in kilograms produced per inhabitant per day by households in the city's communes. Calculations were therefore based on the average quantity of garbage produced in each commune, the average population of inhabitants having produced garbage in each commune, and the number of days (2 days) corresponding to the household garbage production period.

The following formula was used to determine the ratio (**Adamou, 2021**):

$$r = Q * P^{-1} * n^{-1}$$

Or :

r: ratio in kg/capita/d

Q: total quantity of household waste in kg

P: reference population having produced the quantity Q of waste per inhabitant

n: number of days corresponding to the production of quantity Q of household waste

Table 3: Waste-to-energy ratios by commune in Kg/hbt/day in Bamako (Adamou, 2021).

Ratios des OM par commune en Kg/hbt/jour			
Communes	Production moyenne d'ordures générées par un ménage dans une commune (en Kg) pendant les 2 jours	Habitants par ménage	Ratios Kg/hbt/jour
Commune I	15,51	24	0,31
Commune II	10,52	17	0,30
Commune III	9,63	20	0,23
Commune IV	14,78	22	0,33
Commune V	9,92	18	0,26
Commune VI	11,16	21	0,26
Moyennes	11,92	20	0,28

2.1.9. Household and similar waste collection system

Collection refers to all the operations involved in gathering waste from its sources of production (homes, transit depots) and transporting it to treatment centers.

The household and similar waste management chain is shown in figure 2.

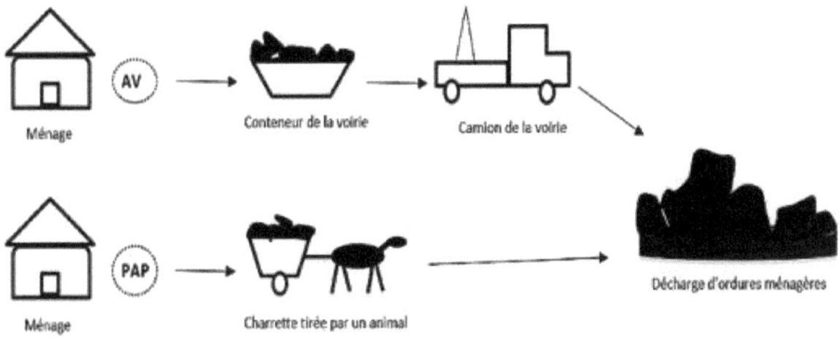

Figure 2: descriptive diagram of the organization of household waste management in Bamako (Adamou, 2021).

2.1.10. Waste management and treatment :

Waste management and treatment is defined as : The collection, transportation, treatment-valorization and disposal of waste and, more broadly, any activity involved in organizing the handling of waste from its production to its final treatment, whether solid, liquid or gaseous, and each has its specific channel and of course the type of waste (Zedam and Laraba, 2019).

In Bamako, sanitation is considered as a process, organized in 3 stages or

These are the "links": the upstream link (waste collection), the intermediate link (evacuation) and the downstream link (treatment, elimination or possible recovery of this waste). (Adamou, 2021).

2.2. Definition of landfill leachate :

Leachate is defined as the water that percolates through the waste, becoming bacteriologically and chemically loaded with mineral and organic substances - "the juice of the garbage" (Djahida, 2016). The composition of landfilled waste, its degree of decomposition, its moisture content and the age of the landfill are the main parameters influencing the composition of leachate. The massive production of leachate generates risks of soil, river and groundwater pollution (Djahida, 2016). Figure 3 shows the components of a water balance

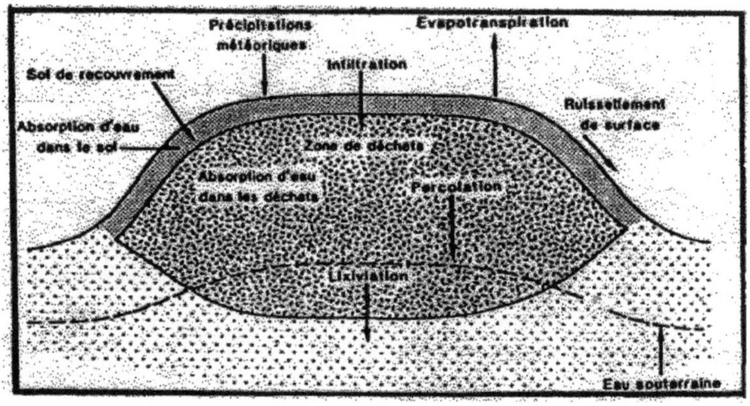

Figure 3: The components of a water balance (Quesnel, 2001).

These phenomena are precipitation, runoff, infiltration and percolation, evaporation and the storage capacity of the soil or waste (Kamissoko and Facourou, 1994). Part of the precipitation is evacuated by surface runoff, another part is returned to the atmosphere by soil and plant evapotranspiration, while the remainder is absorbed by the overlying soil. When the soil's retention capacity exceeds its optimum, the water percolates into the waste. The same phenomenon occurs at waste level: when the moisture content of the waste exceeds its retention capacity, water percolates into the underlying soil and then reaches the collection system in the form of leachate (Quesnel, 2001).

2.3 Political and administrative organization

The District was, according to Ordinance n°78-32/CMLN of August 18, 1978, both an administrative division of the State, located at the same hierarchical level as the region, and a decentralized collectivity endowed with legal personality and financial autonomy (Diakité et *al*, 2018).

Legal framework

The preamble to the constitution states that the sovereign people are committed to improving the quality of life and protecting the environment and cultural heritage. Article 15 further specifies that everyone has the right to a healthy environment. Protecting the environment, defending the environment and promoting the quality of life are a duty for everyone and for the State (Diakité et *al*, 2018).

Specific legislation has been passed to regulate the sector:

➢ Order 69/DB of June 19, 1990 creating the DSUVA ;

➢ Arrêté N°70/DB du 30 Aout 1989 portant création de la Brigade urbaine de Protection de l'Environnement (BUPE) ;

➢ Circular letter No. 0010/DB-GD of March 9, 1993 on the regulation of the activities of collection organizations in the District of Bamako;

➢ Law 95-034 AN-RM of April 12, 1995 on the local authority code, amended by Law N°98-010 of June 19, 1998, defines the powers of local authorities in environmental matters;

➢ Law 96-025 of February 21, 1996 on the special status of the District of Bamako and its communes;

➢ Decree N°03-594/P-RM of December 31, 2003 on Environmental Impact Assessments (EIA)

➢ Arrêté N°1/CD-GD du 23 janvier 1986 portant institution des plans types de puisards et latrines dans le périmètre du District de Bamako ;

➢ Arrêté N°22/CD-GD du 10 juin 1982 portant règlement de police en matière d'hygiène publique et d'assainissement dans le périmètre du District de Bamako;

➢ The adoption in 2008 of a new text on the organization, operation and organic framework of the DNACPN and its decentralized departments (DRACPN, SACPN).

➢ Decree No. 2014-0572/P-RM of July 22, 2014 setting out the details of the powers transferred from the State to local authorities in the field of sanitation and the fight against pollution and nuisances.

In addition to the texts of sanitation relies on policies and strategies (Diakité et *al*, 2018):

- ❖ The National Solid Waste Strategy,
- ❖ The National Strategy for Special Waste,
- ❖ The National Strategy for the Transfer of Sanitation Skills.

National legislation

The Malian Constitution

The Constitution of February 25, 1992 demonstrated Mali's determination to protect the environment and the quality of life. It created a citizen's right to a healthy environment, and made environmental protection a duty for all citizens, as well as for the State.

Law n°01-020/AN-RM of May 30, 2001 on pollution and nuisances.

Article 16: Any natural or legal person, public or private, carrying out an activity which is a source of pollution or which may present a danger to water resources and environmental hygiene, must take all appropriate measures to curb or prevent the danger identified or suspected.

All polluters must bear the costs of their polluting activities.

Law n°96-050/AN-RM on the principles governing the establishment and management of local authority property.

Article 11: Local authorities are responsible for managing, developing, conserving and safeguarding the ecological balance of their land.

To this end, they will draw up a land-use plan that will specify in particular: forestry; agriculture; pastoral; wildlife; fish farming; mining; housing.

Decree no. 01-397/P-RM of September 06, 2001 setting out the procedures for managing atmospheric pollutants.

Article 25: Any motorized machine whose operation generates gases, fumes or other substances in the atmosphere that are likely to inconvenience the public or compromise the environment, public health and safety is subject to the provisions of this decree.

Decree n°01-394/P-RM of September 06, 2001 setting out solid waste management procedures.

Article 4: Any person whose activity produces solid waste or who holds it in conditions likely to harm human or animal health, to have negative effects on the soil, flora or fauna, to cause degradation of sites and landscapes or to pollute the air or water, or to generate odors or visual nuisances, is required to dispose of it in accordance with the provisions of this decree.

Decree no. 08-346/P-RM of June 26, 2008 on environmental and social impact assessments.

Article 2: The purpose of the environmental and social impact study is to :

the prevention of environmental degradation and deterioration in the living conditions of local populations as a result of project implementation; the reduction and/or repair of damage caused to the environment through the application of measures to mitigate, compensate for or correct the harmful effects of project implementation; the optimization of the balance between economic, social and environmental development; the participation of the populations and organizations concerned in the various phases of projects; the provision of information necessary for decision-making.

Decree no. 07 135 P-RM of April 16, 2007 establishing the list of hazardous wastes.

Article 1: The following wastes are classified as hazardous waste: substances and articles containing or contaminated by polychlorinated biphenyls (PCBs), polychlorinated terphenyls (PCTs) and wastes containing beryllium and its compounds; hexavalent chromium compounds; cadmium and its compounds; mercury and its compounds; lead and its compounds.

Article 2: The import, storage, transit, offer or sale, acquisition or transfer for valuable consideration or free of charge, possession, transformation, destruction, neutralization and disposal of hazardous waste from other countries are prohibited and punishable in accordance with the legislation in force.

Law n°02-006/AN-RM of January 31, 2002 on the water code.

Article 16: Any natural or legal person, public or private, carrying out an activity which is a source of pollution or which may present a danger to water resources and environmental hygiene, must take all appropriate measures to curb or prevent the danger identified or suspected.

All polluters must bear the costs of their polluting activities.

International legislation

In addition to national legislation, Mali has signed and ratified several international conventions, some of which have an impact on environmental quality management:

-**The Basel Convention** on the Control of Transboundary Movements of Hazardous Wastes and their Disposal, Aware of the damage that hazardous wastes and other wastes, as well as transboundary movements of such wastes, can cause to human health and the environment, this convention establishes the measures to be taken to ensure that the management of hazardous wastes, including their transboundary movement and disposal, is compatible with the protection of human health and the environment. The transfer of hazardous waste from Mali to other countries requires notification in accordance with the provisions of the Convention. Disposal must be carried out in an environmentally sound manner, i.e. in compliance with discharge standards.

Adopted in Basel on March 22, 1989, this Convention was ratified by Mali on September 15, 2000 (regulating the import and export of hazardous wastes, their recovery and disposal in an environmentally sound manner). Dating back to 1992 and with its secretariat in Geneva, it now has 168 member states. What happened in Côte d'Ivoire, with the import and illegal dumping of toxic waste, showed that there is still much to be done in this area. Many countries have neither the necessary control mechanisms nor the technical facilities to dispose of hazardous waste without endangering human health and the environment. This Convention remains without a financing mechanism (Maiga, 2018).

The Basel Convention sets out a list of categories of supposedly hazardous waste (biomedical waste, PCBs, etc.) and a list of wastes with clearly identified constituents (mercury, lead and asbestos). In view of these characteristics, it can be said that atmospheric pollutants fall within the scope of the Convention. For its implementation, each party can exercise its right to prohibit the import of hazardous waste and decide to halt unauthorized waste movements. On September 22, 1995, an amendment to the Convention was adopted, to prohibit all shipments of hazardous waste from OECD countries to non-OECD countries, so that the latter can carry out final disposal. This amendment was also to prohibit, from January 1, 1998, all such shipments destined for recovery, recycling, reuse, remediation or other alternative uses (Diop et al, 2014).

-The Bamako Convention on the Ban of the Import into Africa and the Control of Transboundary Movements of Hazardous Wastes was adopted in Bamako on January 31, 1991 and ratified by Mali on February 21, 1996. Like the Basel Convention, the Bamako Convention helps to prevent the illegal traffic of hazardous waste across the African continent. It recommends the use of environmentally sound technologies for the disposal of hazardous waste.

This Convention has a regional vocation, and its scope of application is sometimes broader than that of the Basel Convention.

The OAU saw the need for an agreement to regulate the transboundary movement of hazardous waste in Africa. Indeed, the wording of the Basel Convention did not seem sufficiently clear and firm to meet the expectations of the members of the Organization of African Unity (OAU). Signed on January 30, 1991 and coming into force on April 22, 1998, the Bamako Convention prohibits the import into Africa of hazardous and radioactive waste from non-contracting parties. It therefore subjects movements within the African continent to a system close to that defined by the Basel Convention procedures (Maiga, 2018).

III. MATERIALS AND METHODS

3.1 Hardware

3.1.1. Study site :

Located between 12°38' north latitude and 7°59' west longitude, the District of Bamako covers an area of approximately 267 km^2 and extends 22 km from west to east and 12 km from north to south, on either side of the Niger River. It is divided into two main banks and has a Sudanian climate. Its population is estimated at 2384780 (DRPSIAP-DB, 2014), representing a density of 8,932 inhabitants per km^2. The city produces around 1,400 tonnes of household waste per day, or 0.6 kg per inhabitant (Mairie du District de Bamako, 2014) (Outtara et *al*, 2017).

Table 4: Breakdown of resident population and households by commune.

District of Bamako	Number of Households	Number Of men	Number of Women	Population Total
Commune I	52 943	168 348	167 059	335 407
Commune II	25 532	80 272	79 533	159 805
Commune III	19 977	63 854	65 018	128 872
Commune IV	49 210	149 840	150 245	300 085
Commune V	64 078	208 075	206 593	414 668
Commune VI	76 436	238 506	231 763	470 269
Total	**288 176**	**908 895**	**900 211**	**1 809 106**

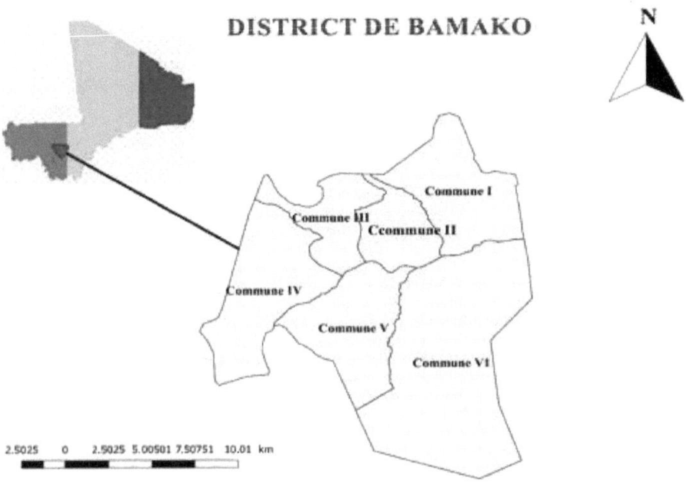

Figure 4: Map of Bamako district (Doucoure, 2023)

The Bamako District is underlain by gneissic granite and shale covered by sandstone sediments. The upper part of these sediments is fissured, while the unfractured underlying part seems to constitute an impermeable formation representing the base or substratum of the aquifer (Aboubacrine, 1991).

The river has cut more or less deeply into the schists and granitoids of the basement and the sedimentary cover. Its alluvial deposits occupy the river's major bed; recent deposits fill in the depressions in the riverbed after each flood.

Two types of surface formation are distinguished (Diakité et *al*, 2018):

- Soils resulting from rock weathering and lateritization.
- Alluvial formations occupying the major and minor beds of the river and its tributaries.

In Bamako, the dry season lasts from November to April, and the rainy season from May to October. The monthly variation in the number of rainy days ranges from 0 to 24 days, with the maximum occurring in August. Annual rainfall varies from 722.6 mm to 1499.5 mm from1936 to 1991, with an average of 1100mm (Diakité et *al*, 2018).

Despite being located in the northern Sudanian zone, Bamako recorded 1,499.5 mm of rain in 1967. However, recent decades of desertification have resulted in a decline in rainfall. The average annual temperature is 27.6°C, with extreme averages of 34.8°C and 21°C (Diakité et *al*, 2018).

3.1.2. Equipment used :

In the course of this study, a variety of equipment was used, including site location, waste characterization, soil analysis, physico-chemical analysis and heavy metal analysis of leachates and groundwater surrounding the landfill:

- A cell phone (Samsung SM-G960W) was used to collect Proget WGS84 GPS data;
- A plastic bottle with a capacity of 330 liters and 1.5 liters was used to store leachate and well water;
- a Burette was used for the various dosages,
- a sartorius TE124S electronic analytical balance was used to weigh samples up to 120g with a display accuracy of 0.0001 g (figure 5);
- A mechanical balance was used to weigh samples over 120 g (figure6);

Figure 5: TE 1245 electronic scale *Figure 6: Mechanical scale*

- A Carbolite Oven was used to dry the soil samples and the soil sorted from the waste (figure 7);

- A NITON XL3t GOLDD++ XRF device (Thermo Scientific) was used to determine the mineral composition of the material (soil and soil sorted from waste) (figure 8);

Figure 7: Carbolite furnace

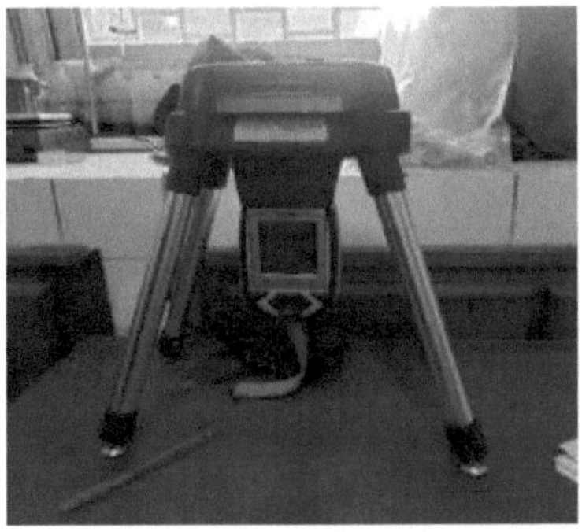

Figure 8: Niton XRF Analyzer

- A HANNA HI2209 pH meter was used to measure pH (figure 9);
- A HANNA HI8633 conductivity meter was used to calculate conductivity (figure 10);

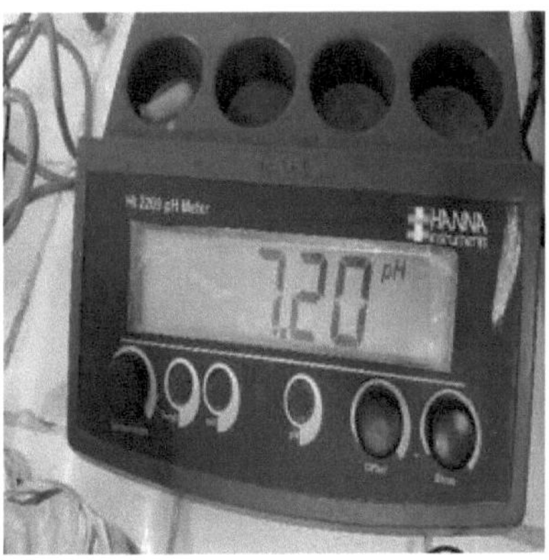

Figure 9: HI2209 PH meter (HANNA) (HANNA)

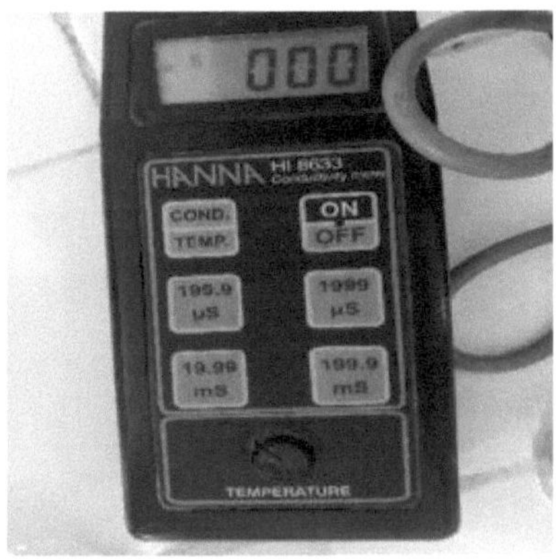

Figure 10: HI8633 conductivity meter

- A HANNA HI 98703 turbidity meter was used to calculate turbidity (figure 11);
- A UV-VIS Spectrophotometer was used to calculate certain physico-chemical compounds (TA, $SO4^{2-}$, NO^{3-}, NO^{2-}) (figure 12);

Figure 11: Turbidimeter HI 98703 (HANNA) *Figure 12: UV- Spectrophotometry*

- An Agilent Technologies 4210 MP-AES flame spectrophotometer was used to analyze leachate and well water for heavy metals (F, Pb, Cd, Se, Si, K, As, Zn, etc.) (figure 13);

Figure 13: Agilent Technologies 4210 flame spectrophotometer

- Puisettes were used to draw water from the wells (figure 14);

Figure 14: Catch

- And a few small items of laboratory equipment: picette, filter paper, glove, bulb, vat, beaker, test tube, pipette, etc. were also used.

3.1.3 : Reagents :

The reagents used are :

- HCl, Phenolpahtalein 1%, Helianthine, these reagents were used for TA and TAC analyses;
- EDTA, Sodium Hydroxide Solution and Ammoniacal Buffer pH 10, were used for Total Hardness (Mg^{2+}, Ca^{2+}) and Calcium (Ca^{2+}) analyses;
- Silver nitrate solution 0.025mo/L, potassium chromate or potassium bicromate 10% were used for Chloride (Cl^-) analysis;
- HCl 10%, Barium chloride were used for sulfate ion (SO_4^{2-}) analysis;
- Pure ammonia, zambeli reagent, 0.5% sodium salicylate solution, 1% NaCl and sodium-potassium double tartate solution were used for nitrite (NO_2^-) and nitrate (NO_3^-) analyses;
- Hydroxylamine chloride 10% and Phenanthroline 0.42%, used for Fe^{2+} analysis

3.2 Method :

3.2.1. Literature search

Documentary research consisted of searching and consulting books, theses, university dissertations, reports and scientific articles related to the theme. We also took advantage of Internet sources. This work enabled us to gain a better understanding of the subject. To make an inventory of the various transit depots (authorized and unauthorized) in the Bamako district, and to characterize them. Among the various transit depots in Bamako, we sampled three official depots (Medina-coura, Dar-salam and Lafiabougou) and one unauthorized depot (Badalabougou).

Site selection

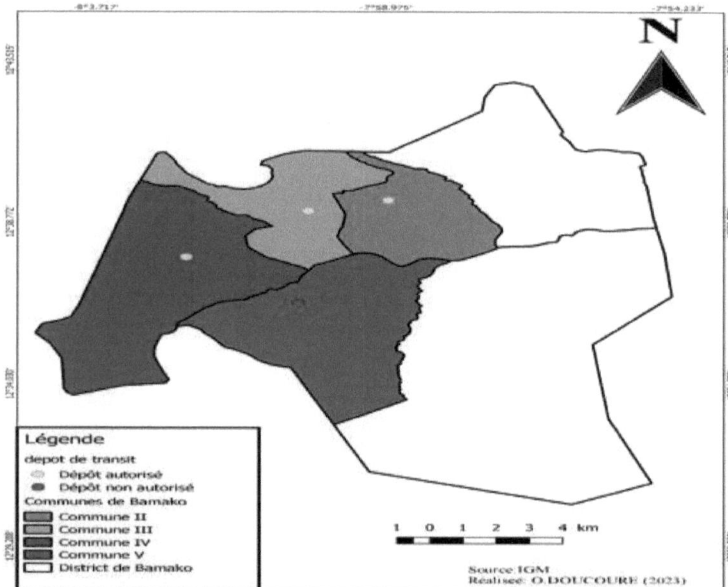

Figure 15: Transit depot sites in the various communes of the Bamako district (by Ousmane Doucoure 17/03/2023)

3.2.2 Inventory of different transit depots in Bamako:

Commune I: The commune lacks official transit depots. However, it does have two unauthorized transit depots: Banconi and Doumanzana. Thirty-three (33) MSEs collect the commune's waste daily with on-board means to converge on the CFP transit depot (Diakité et *al*, 2018).

Commune II: The only transit depot is at CFP, which receives most of the waste from several districts in communes 1 and 2. Every day, it is literally invaded by commercial waste (agricultural and household products). Because it is so busy, it is in an advanced state of disrepair. Fifty-three (53) formal MSEs are counted every day around the depot, including 23 for commune II (Diakité et *al*, 2018).
Figure 16 shows the satellite image of the Médina-coura landfill site

Figure 16: Satellite image of Médina-coura landfill, Google Earth, images from 01/01/2023

Commune III: The only transit depot in Dar Salam is well equipped and complies with standards in terms of garbage dwell time. A second transit depot in Samé is equipped but not operational. The number of MSEs working with the mayor's office is eleven (11) for around twenty neighborhoods (Diakité et *al*, 2018).
Figure 17 shows satellite image of the Dar-salam landfill.

Figure 17: Satellite image of Dar-salam landfill, Google Earth, images from 01/01/2023

Commune IV: Twenty-eight (28) MSEs work with the mayor's office to manage the commune's waste, but as several groups are no longer operational, they are unable to meet the commune's waste disposal needs. The town has four (04) authorized transit depots (LAFIABOUGOU-cimetière, le luxambourg, Sébéninkoro -flanc colline, Kalabambougou), the largest of which, LAFIABOUGOU- CIMETIERE, is currently under construction and handles 500,000 m3 of waste a day. Today, all attention is focused on this site (Diakité et *al*, 2018).Figure 18 shows a satellite image of the Lafiabougou landfill.

Figure 18: Satellite image of Lafiabougou landfill, Google Earth, images from 01/01/2023.

Commune V: As in other communes, sanitation remains a major concern for the population. Waste management is mainly handled by MSEs involved in pre-collection in the various neighborhoods, but their equipment does not meet the commune's needs. There are 45 MSEs working on the basis of communal zoning. The commune lacks authorized transit depots (Diakité et *al*, 2018). Figure 19 illustrates the Badalabougou landfill satellite image.

Figure 19: Badalabougou landfill satellite image, Google Earth, images from 01/01/2023.

Commune VI: The commune has no transit depots, and the many anarchic depots continue to be fed. The 29 more or less functional MSEs do not cover waste disposal needs. There is a lack of transit depots, and existing depots are neither equipped nor authorized. Of 52 existing depots, only Dianeguela was authorized. This site was annexed by the construction of the Palais des Pionniers after 52,000 m3 of waste was evacuated there (DNACPN) thanks to the combined efforts of the Ministry in charge of the Environment and the District Town Hall (Diakité et *al*, 2018).

3.2.3. Sampling :

Our sampling consisted of sampling at the various transit depots (authorized and unauthorized). During our sampling process, samples were drawn at random from the three authorized transit depots (Medina-coura, Dar-Salam, Lafiabougou) and one unauthorized depot located on Badalabougou hill. For each dump, we obtained the Proget WGS84 GPS coordinates.

The results are shown in the following table:

Table 5: Geographical coordinates of the four landfills sampled.

Municipality	Transit depots	Latitude	Longitude	Altitude in (m)
Commune II	Medina-Coura CFP	12.65794348	-7.99282225	326.32

Commune III	Dar-Salam	12.65035323	-8.00700090	337.43
Commune IV	Lafiabougou	12.62227620	-8.03873028	368.67
Commune V	Badalabougou	12.60689359	-7.99210088	411.88

3.2.3.1.Solid waste and soil sampling at landfill sites

Our solid waste and soil sampling phase began on Saturday, October 5, 2022 at 10 am. During this phase, we manually collected around 2 kg of solid waste and 2 kg of soil in plastic bags. The soil samples were taken near the landfill from the ground surface (above natural ground level) to a depth of 30 cm. By sorting the household waste, we were able to obtain the actual characteristics of the raw state of the waste.

The samples taken were analyzed at the FST/USTTB measurement center.

- Solid waste was sorted by category. Each category was weighed using mechanical and electronic scales, and the data recorded in the notebook. The items sorted by category are shown in Table 7 ;
- The soils were analyzed for the composition of contaminating heavy metals using a device called XRF.

Figure 20: Image of soil samples taken; (photo, credit Doucoure)

Figure 21: Image of waste samples taken; (photo, credit Doucoure)

3.2.3.2. Collection of leachate and well water

Leachate and groundwater sampling was carried out manually. During this sampling process, both leachate and well water were sampled. Approximately one liter of leachate was collected from the landfill sites, and one liter from the wells surrounding the landfill sites. The leachate and well water samples were taken in polyethylene bottles of 330 ml and 1500 ml volume. Well water was sampled using well catch basins placed on the well curbstones. A total of 22 wells were sampled.

Figure 22: Leachate samples taken. *Figure 23*: well samples taken.

Table 6: Geographical coordinates of sampled wells surrounding landfills.

Transit depot	Samples	Latitude	Longitude	Altitude (in m)	Waste shaft distances in (m)
Medina-coura	PM1	12,65740962	-7,99275774	372,87	20
	PM2	12,65815061	-7,99278586	369,4	5
	PM3	12,65744154	-7,99208820	376,23	9
	PM4	12,65722263	-7,99165104	360,79	18
	PM5	12,65761164	-7,99266508	360,32	10
	PM6	12,65579170	-7,98946723	368,35	328
Dar Salam	PD1	12,65052217	-8,00712196	367,41	17
	PD2	12,65088543	-800687359	377,37	43
	PD3	12,65077924	-8,00679113	368,68	50
	PD4	12,65064933	-8,00649134	367,37	48
Lafiabougou	PL1	12,62222052	-8,03781434	365,02	46
	PL2	12,62292463	-8,03801369	359,22	7
	PL3	12,62291478	-8,03815627	356,11	17

	PL4	12,623245 46	- 8,03942100	374,23	13
	PL5	12,623262 52	- 8,03968434	367,98	15
	PL6	12,621714 05	- 8,03872237	397,69	70
	PL7	12,621783 46	- 8,03767082	363,13	103
	PL8	12,626061 30	- 8,03759041	385,02	326
Badalabougou	FB1	12,606893 59	- 7,99210088	411,88	184
	PB1	12,612464 09	- 7,99484495	375,33	530
	PB2	12,613658 32	- 7,99362723	358,64	588
	FB2	12,608840 32	- 7,99122945	403,82	80

Figures 24, 25, 26 and 27 show the points of the discharge sites and wells sampled in the various communes (Medina-coura, Dar-salam, Lafiabougou and Badalabougou).

Figure 24: Wells sampled near the Medina-coura CFP transit depots, (Google Earth 02/01/2023

Figure 25Wells sampled near the Dar-salam transit depots, Google, (Earth 02/01/2023).

Figure 26: Wells sampled near the Lafiabougou transit depots, Google (Earth 02/01/2023)

Figure 27Wells and boreholes sampled near the Badalabougou transit depots, Google (Earth from 02/01/2023)

- Sites sampled in transit depots.

- Wells sampled near transit depots.

3.2.4. Sample analysis

3.2.4.1. Physical characteristics of landfill waste

After drying the samples in a carbolite oven at 105°C for 24 hours, the soil was analyzed using an XRF device.

3.2.4.2. Physico-chemical parameters of leachates and groundwater near transit depots in the Bamako district

❖ **Physical parameter**

a. pH: Hydrogen potential

The pH (hydrogen potential) is one of the fundamental characteristics of water. The pH is a measure of acidity, i.e. the concentration of hydrogen ions (H^+), or basicity, i.e. the OH^- ions contained in the substance. These parameters were measured using a Hanna-type pH meter.

b. Conductivity (CE) :

The conductivity of water reflects its ability to carry electric current. Most of the substances dissolved in water are in the form of electrically-charged ions, so measuring conductivity enables us to assess the quantity of salts dissolved in the water. It is therefore logical to assume that the higher the mineralization, the greater the conductivity of the water. The measurement was determined using a Hanna-type conductivity meter.

c. Turbidity :

Turbidity reduces the transparency of a liquid. It therefore reflects the presence of particles suspended in the water (organic debris, clays, micro-organisms, etc.). Turbidity was measured using a Hanna Turbidimeter.

❖ **Chemical parameter :**

a. Alkalimetric Titration (AT) and Total Alkalimetric Titration (TAC):

Water alkalinity corresponds to the presence of bicarbonate, carbonate and hydroxide ions. Assays were carried out by potentiometer and volumetry.

b. total hardness TH ($Ca^{2+}Mg^{2+}$) :

Water hardness is the sum of metal cation concentrations, excluding alkali metals (Na^+ , K^+) and H+. In most cases, it is mainly due to Ca^{++} and Mg^{++} ions. The presence of these two cations in water often tends to reduce the toxicity of metals, but their excess prevents the formation of foam with soap. Hardness is measured in mg CaCO3 per liter. Measurements were made by potentiometer and volumetry.

c. Determination of sulfate ions: SO_4^{2-}

These elements were determined by UV Visible Spectrometry.

d. Determination of Nitrite ions (NO_2^-) :

Nitrites (NO^-_2) are undesirable compounds in the composition of drinking water. Its presence may be due to the incomplete oxidation of NH_4 or the reduction of NO_3^- present in the water. These elements were determined by UV Visible spectrometry.

e. Determination of Nitrate ions: NO_3^-

$NO^{(3-)}$ ions are generally the consequence of nitrification of organic nitrogen or come from the dissolution of fertilizers. These elements were determined by UV Visible spectrometry.

- **Heavy metals (Pb, Cr, Iron, Zn and Cu etc.)**

Parameters were determined by flame spectrophotometer (atomic absorption).

NB: For the leachate analyses, we diluted the samples to one-tenth (1/10).

IV. RESULTS AND DISCUSSION

4.1. Results

4.1.1. Physical composition of landfill waste

The results of sorting waste taken from the various landfills in the Bamako district are shown in Table 7 below.

The results in Table 7 show that wood is the waste category most present in Bamako landfills, with an average of 347.75 g and a percentage of . After wood, plastics come second with an average of 341.64 g and a percentage of .

Table 7: Quantities of various elements present in household waste from landfills in grams (g).

Samples	Medina-coura	Dar-salam	Lafiabougou	Badalabougou	Average
Plastics	400	211,51	335,05	420	341,64
Wood	400	321	390	280	347,7
Fabrics	450	84,15	7,01	120	165,29
Floors	220	100	800	200	330
Pebbles	70	11,06	18,56	15,96	28,895
Glasses	300	0	4,27	0	152,135
Bones	74	1,23	0	0	37,615
Papers	400	100	300	90	222,5
Hair	360	0	3,04	0	181,52
Scrap metal	590		0	20	305
Egg	0	12,73	0	0	12,73
dead insects	0	0,68	0	0	0,68
Feathers	0	4,04	0	0	4,04
Aluminum foil 0		14,25	0	20,5	17,37
Rubber	0	1,68	0	90	45,84
Amounts	3364	831,33	1657,93	1045,96	1178,40

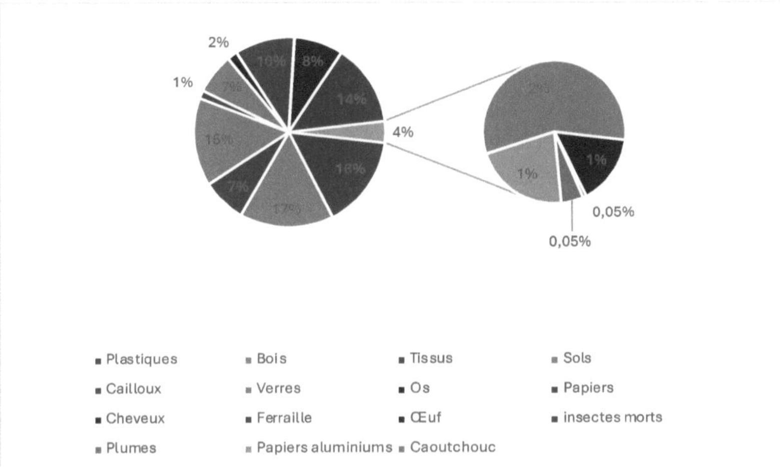

Figure 28: Average percentage composition of household waste in the Bamako district.

4.1.2. Characteristics of heavy metals in soil and sorted waste soil

Table 8 shows the mineralogical composition of major elements in soil and sorted soil from waste at the various transit depots. The table shows a high load of Fe_2O_3 and CaO with Fe_2O_3 values ranging from 1.98% to 5.and an average of 3.CaO with values ranging from 0.79% to 6.65% with an average of 8.45%

Table 9 shows the trace element content of the soil studied. Chromium (Cr), zinc (Zn), cadmium (Cd), lead Pb) and copper (Cu) levels are higher than in standard and ordinary . This explains the presence of trace metallic elements in the soil. Analysis of these values shows that waste is one of the main sources of heavy metal contamination in soil. Unsorted waste can infiltrate the soil through landfill and contaminate the water table.

Table 8: Mineralogical composition of major elements in soil and sorted soil of waste from various transit depots as a percentage (%)

Elements	Unit	SM	SdM	SD	SdD	SL	SdL	SB	SdB	Ordinary floor standards
Fe_2O_3	%	**4,23**	**2,32**	**1,98**	**1,71**	**3,59**	**4,22**	**5,38**	**5,03**	**1,085**
CaO	%	**6,65**	**8,77**	**4,61**	**20,92**	**4,65**	**5,62**	**7,88**	**8,51**	**1,91**
K_2O	%	1,15	0,79	1,26	**2,30**	0,53	1,07	1,17	1,15	**2**
SiO_2	%	11,78	11,03	12,23	5,16	15,99	11,54	11,30	7,62	
Al_2O	%	0,64	0,77	0,85	0,26	0,63	0,52	0,71	0,74	
MgO	%	0	0	0	0	0	0	0	0	**0,83**
Cr_2O_3	%	0,023	0,014	0,0073	0,0064	0,014	0,034	0,017	0,011	
P_2O_5	%	0,27	0,34	0,054	0,64	0,17	0,24	0,36	0,32	
TiO2	%	0,12	0,11	0,21	0,08	0,09	0,10	0,17	0,13	

Table 9: Mineralogical composition of minor elements in soil and sorted soil of waste from the various transit depots in (ppm).

Elements	**Unit**	**SM1**	**SdM**	**SD**	**SdD**	**SL**	**SdL**	**SB**	**SdB**	**Ordinary floor standards**
S	**Ppm**	2384,83	1748,46	87,31	985,4	257,97	1485,79	1493,46	1059,6	
Cl	**Ppm**	1415,83	304,06	643,81	969,09	0	1784,39	1236,44	860,51	
V	**Ppm**	46,92	40,08	32,78	0	48,03	55,85	75,08	32,58	
Cr	**Ppm**	**161,21**	98,7	50,15	44,03	96,95	**234,79**	**123,26**	76,99	**100**
Mn	**Ppm**	200,98	102,43	252,96	637,26	0	211,23	324,76	0	
Cu	**Ppm**	**56,8**	23,18	0	**32,25**	0	0	**124,27**	**134,72**	**30**
Zn	**Ppm**	**354,68**	**234,91**	**93,9**	**214,33**	**91,36**	**111,53**	**403,77**	**402,94**	**50**
As	**Ppm**	4,05	0	0	0	0	3,82	4,49	4,21	
Visit	**Ppm**	1	1	1	1	1	1	1	1	
Rb	**Ppm**	31,47	11,75	20,29	25,93	7,35	12,34	15,58	15,42	
Sr	**Ppm**	233,72	173,09	146,26	471,44	89,74	108,32	194,58	178,26	
Zr	**Ppm**	173,71	105,52	367,31	231,51	98,61	123,92	111,69	124,92	

Nb	**Ppm**	7,36	2,92	14,99	7,52	3,43	3,77	4,55	3,98	
Mo	**Ppm**	1	1	1	1	1	1	1	1	
Ag	**Ppm**	5,95	0	0	0	0	3,42	3,71	0	
Cd	**Ppm**	5,72	0	6,72	0	0	0	0	0	
Sn	**Ppm**	0	0	14,06	0	0	0	0	0	
Sb	**Ppm**	0	0	9,64	0	0	0	10,78	0	
Ba	**Ppm**	438	206,64	147,02	579,05	193,04	257,29	216,97	255,34	
Pb	**Ppm**	**42,11**	**14,71**	6,01	7,14	0	**17,11**	**34,36**	**43,26**	**10**

NB: Values in bold exceed the standard.

4.1.3. Physico-chemical parameters of leachates

Table 10 shows the values of physicochemical parameters for leachates from landfills (Medina-coura, Dar-salam, Lafiabougou and Badalabougou). The results of physico-chemical analyses are presented in the table below.

In Table 10, the results show that the pH of the leachates ranges from 7.33 to 7.96, with an average of 7.59. This is within the acceptable reference values. This is within the acceptable reference limits of 3.7 to 8.8, which means that landfill leachates collected from the various transit depots in the Bamako district are slightly neutral to alkaline in terms of hydrogen potential (pH). However, leachate conductivity levels are high, fluctuating between 16200 and 26000 μS/cm, with an average of 21950 μS/cm. This is well above the Malian wastewater discharge standard of 2,500 μS/cm. In addition, leachate turbidity exceeds 1000 NTU. Sulfate levels range from 24.92 to 226.72 mg/L, with an average of 147.39 mg/L. The nitrate concentration recorded in the leachates is well above the maximum Malian wastewater discharge standard of 30 mg/L, ranging from 40.81 to 80.48 mg/L, with an average of 64.02 mg/L. On the other hand, nitrite concentration is low, with an average value of 0.013 mg/L. In terms of leachate sulfate content, only the Badalabougou landfill is very high, with a value of 2492.24 mg/L. The Malian standard for sulfate discharge is set at 1000 mg/L. In addition, the leachates contain significant quantities of heavy metals, with Cu, Pb and Al values exceeding the accepted standard for wastewater discharge in Mali.

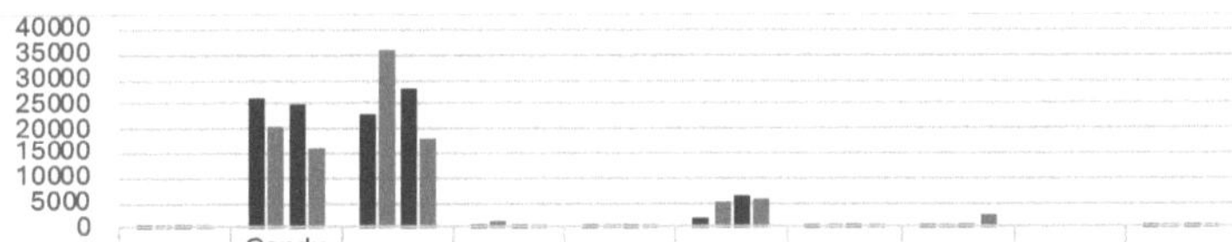

	pH	Conductivité	TAC	Mg2+	Ca2+	Cl-	Fe2+	SO4-	NO2-	NO3-
LixMC	7,33	26000	23000	926	720	1597,5	2,24	226,72	0,002	40,81
LixDS	7,5	20600	36000	1056	800	5325	5,89	121,55	0,016	80,48
LixLa	7,57	25000	28000	672	640	6212,5	3,23	216,37	0,022	66,4
LixBa	7,96	16200	18000	576	480	6035	4,9	2492,24	0,013	68,41

LixMC LixDS LixLa LixBa

Figure 29: Variation in leachates from landfills (Medina-coura, Dar-salam, Lafiabougou, Badalabougou).

Table 10: Physico-chemical parameters of landfill leachates (Medina-coura, Dar-salam, Lafiabougou, Badalabougou)

Parameter	**Unit**	**Lix Medina-coura**	**Lix Dar-salam**	**Lix Lafiabougou**	**Lix Badalabougou**	**Malian discharge standard**
pH	-	7,33	7,5	7,57	7,96	**6.5 - 9.5**
Conductivity	µS/cm	**26000**	**20600**	**25000**	**16200**	2500
Turbidity	NTU	>1000	>1000	>1000	>1000	
TA	-	0	0	0	0	
TAC	-	23000	36000	28000	18000	
$Mg2^+$	mg/L	926	1056	672	576	
$Ca2^+$	mg/L	720	800	640	480	
Cl^-	mg/L	**1597,5**	**5325**	**6212,5**	**6035**	1200
Fe^{2+}	mg/L	**2,24**	**5,89**	**3,23**	**4,9**	2
$SO4^-$	mg/L	226,72	121,55	216,37	24,92	1000
$NO2^-$	mg/L	0,002	0,016	0,022	0,013	0,6

NO3^{-}	mg/L	**40,81**	**80,48**	**66,4**	**68,41**	30
Visit	mg/L	0	1,29	1,36	0	
Zn	mg/L	0,5	0,46	0,73	0,51	
Cd	mg/L	0	0	0	0	0,5
Sr	mg/L	0,49	1,2	1,4	0,62	
Ba	mg/L	0,52	0,55	0,67	0,39	
Cu	mg/L	0,05	**0,15**	**0,15**	**0,35**	0,1
Ni	mg/L	0,09	0,48	0,89	0,64	2
As	mg/L	0	0	0	0	0,5
Pb	mg/L	**0,31**	**1,15**	**1,78**	**1,1**	0,2
Mn	mg/L	0,31	0,88	0,83	0,98	2
Cr	mg/L	0,07	0,42	0,26	1,1	2
Al	mg/L	0,11	**1,09**	**1,24**	**2,96**	1

4.1.4. Physico-chemical parameters of wells collected from landfill sites

Tables 11, 12, 13, 14, 15, 16, 17 and 18 show the results of physico-chemical and metal analyses of the wells surrounding the various landfills in Bamako.

- **Medina-coura :**

The results of Table 11 show that the pH values of the Medina-Coura wells are slightly acidic, ranging from 5.8 to 6.75, with an average of 6.2. Pml has a slightly high conductivity, with a value of 829, which is close to the Malian standard of 1500 mg/L compared with the other wells. Turbidity at pm2 was higher than the norm, at 11 NTU. Nitrite concentrations in pm2, pm3, pm4 and pm5 are very high, with values ranging from 0.39 to 1.43 mg/L. These far exceed the accepted Malian standard of 0.2 mg/L. Table 12 shows heavy metal concentrations in well water. The analysis revealed that selenium contamination levels in the wells ranged from 2.55 to 3.75 mg/L, with an average of 3.22 mg/L, well above the Malian standard of 0.01 mg/L. Arsenic concentrations were also high, with values ranging from 2.88 to 5.3 mg/L and an average of 3.61 mg/L, respectively exceeding the 0.01 standard. Pml and Pm6 are very high in lead, with values ranging from 0.23 to 0.32.

Table 11: Physico-chemical parameters of the Medina-coura well (commune II)

Paramete r	Unit	Pml	Pm2	Pm3	Pm4	Pm5	Pm6
pH	-	6,75	6,24	6,18	5,8	6,47	6,57
Conductiv ity	µS/cm	892	3	2	108	851	320
Turbidity	NTU	5	**11**	3	1	1	4
TA	-	0	0	0	0	0	0
TAC	-	620	360	220	100	680	215
Mg^{2+}	mg/L	23,04	18,24	7,68	3,84	25,92	27,54
Ca^{2+}	mg/L	67,2	46,4	17,6	8	49,6	57,66
Cl^-	mg/L	79,87	47,92	49,7	51,47	111,82	62,11
Fe^{2+}	mg/L	0,011	0,007	0,001	0,034	0	0,002
SO_4^{2-}	mg/L	10,78	1,21	0	0	0	0

NO_2^-	mg/L	0	**1,43**	**1,38**	**0,67**	**0,39**	0,02
NO_3^-	mg/L	1,93	4,76	4,44	14,5	3,84	16,55

Table 12: Heavy metal parameters for Medina-coura wells

Elements	Unit	Pm1	Pm2	Pm3	Pm4	Pm5	Pm6
Visit	mg/L	**2,55**	**3,15**	**2,81**	**3,44**	**3,66**	**3,75**
Zn	mg/L	0,03	0,01	0,01	0,03	0,03	0,05
Cd	mg/L	0	0	0	0	0	0,01
Sr	mg/L	0,21	0,11	0,05	0,02	0,15	0,71
Ba	mg/L	0,26	0,05	0,06	0,03	0,18	0,16
Cu	mg/L	0	0	0	0	0	0
Ni	mg/L	**0,19**	**0,06**	**0,08**	0,02	**0,09**	0,19
As	mg/L	**2,88**	**3,03**	**2,61**	**3,57**	**4,3**	**5,3**
Co	mg/L	0,38	0,49	0,05	0,46	0,54	0,13
Pb	mg/L	**0,23**	**0,15**	**0,09**	0	0	**0,32**
K	mg/L	44,58	18,87	16,07	4,34	53,22	98,05
Mo	mg/L	0	0	0	0	0	0
Mn	mg/L	0,03	0	0	0	0,01	0,22
Cr	mg/L	0	0	0	0	0	0
Al	mg/L	0	0	**0,28**	0,07	0,01	0,13

- **Dar-salam:**

Based on the results in Table 13, we can see that the pH of the wells sampled at Dar-salam is slightly acidic, with values ranging from 6.15 to 6.6 and an average of 6.37. The nitrate concentration of pd1, pd2, is a little high, with values ranging from 25.05 to 35.15 mg/L. The latter is slightly closer to the accepted Malian standard of 50 mg/L. Table 14 shows the analysis of heavy metals, the results of which indicate contamination of the wells with selenium, arsenic, lead and nickel. With values ranging respectively from 3.29 to 4.09 for Selenium, 0.56 to 6.63 for Arsenic, 0 to 0.09 for Lead and 0.05 to 0.14 for Nickel. The maximum value of the Malian standard is: 0.01 for Se, As and Pb and 0.02 for Ni.

Table 13: Physico-chemical parameters of the Dar-salam well (commune III)

Parameter	Unit	Pd1	Pd2	Pd3	Pd4
pH	-	6,6	6,15	6,36	6,23
Conductivity	µS/cm	399	333	301	315
Turbidity	NTU	1	0	0	0
TA	-	0	0	0	0
TAC	-	200	240	160	213
Mg^{2+}	mg/L	16,32	20,16	17,28	18,21
Ca^{2+}	mg/L	51,2	25,6	44,8	36,7
Cl^-	mg/L	62,12	55,02	79,87	64,09
Fe^{2+}	mg/L	0,05	0	0	0
SO_4^{2-}	mg/L	0	0	0	0
NO_2^-	mg/L	0	0	0	0
NO_3^-	mg/L	32,52	35,15	25,05	21,20

Table 14: Heavy metal parameters for Dar-salam wells

Elements	Unit	Pd1	Pd2	Pd3	Pd4
Visit	mg/L	**3,29**	**3,83**	**3,93**	**4,09**
Zn	mg/L	0,03	0,02	0,04	0,03
Cd	mg/L	0,01	0,01	0	0
Sr	mg/L	0,12	0,06	0,08	0,17
Ba	mg/L	0,15	0,09	0,11	0,35
Cu	mg/L	0	0	0	0
Ni	mg/L	0,09	0,05	0,14	0,1

As	mg/L	**0,56**	**6,35**	**6,63**	**6,2**
Co	mg/L	0	0,55	0,47	0,61
Pb	mg/L	0	0	0,09	0
K	mg/L	0	12,34	20,31	21,26
Mo	mg/L	0	0	0	0
Mn	mg/L	0	0	0,09	0
^{2}Cr	mg/L	0	0	0	0
Al	mg/L	0	0	0,01	0,02

- **Lafiabougou:**

Table 15 shows that the pH of the wells sampled at Lafiabougou are all acidic, with values ranging from 4.23 to 6.28, with an average of 5.4. PL3, PL4 and PL5 have very high conductivities compared with the other wells, which are close to the maximum permitted value of the Malian standard set at 1500 µS/cm. Nitrite concentrations are low, with the exception of well PL1. However, nitrate concentrations in PL3, PL4, PL5 and PL6 are very high compared with the other wells sampled, ranging from 37.22 to 45.97 mg/L, which is close to the maximum permissible standard set at 50 mg/L.
Table 16 shows the results of heavy metal analyses. Analysis of this table shows that the wells sampled in the vicinity of the Lafiabougou landfill are very rich in heavy metals. Selenium (Se), arsenic (As) and lead (Pb) are well above the maximum Malian standard of 0.01 mg/L. The values of these metals range respectively from 3.6 to 4.32 mg/L for selenium, 0.1 to 7.19 mg/L for arsenic, and 0 to 0.61 mg/L for lead. Wells PL2, PL3, PL4 and PL5 are very rich in Aluminum (Al), with values ranging from 0.27 to 0.44 mg/L, exceeding the maximum permitted standard of 0.2 mg/L. The wells also contain traces of Nickel and Lead, with values ranging from 0.1 to 0.23 mg/L for Ni and 0 to 0.61 for Pb.

Table 15: Physico-chemical parameters of Lafiabougou wells (commune IV)

Parameter	Unit	PL1	PL2	PL3	PL4	PL5	PL6	PL7	PL8
pH	-	6,58	5,9	5,58	4,23	4,56	4,86	5,34	6,18

Conductivity	µS/cm	474	786	1010	1488	1111	679	381	321
Turbidity	NTU	2	1	1	0	1	0	1	0
TA	-	0	0	0	0	0	0	0	0
TAC	-	280	120	140	180	100	140	140	110
Mg^{2+}	mg/L	5,76	68,16	52,8	24,96	53,76	25,92	14,4	17,31
Ca^{2+}	mg/L	89,6	48	86,4	78,4	46,4	24	30,4	67,58
Cl^-	mg/L	67,4	221,9	182,8	363,9	246,7	150,9	85,2	77,64
Fe^{2+}	mg/L	0,024	-	0,011	0,037	0	0	0	0,034
SO_4^{2-}	mg/L	7,25	0	0	0	0	0	0	0
NO_2^-	mg/L	**1,23**	0,05	0	0	0,31	0	0	0
NO_3^-	mg/L	27,48	37,1	37,22	45,67	45,97	45,97	25,79	21,18

Table 16: Heavy metal parameters for Lafiabougou wells

Elements	**Unit**	**PL1**	**PL2**	**PL3**	**PL4**	**PL5**	**PL6**	**PL7**	**PL8**
Visit	mg/L	**3,79**	**4,32**	**4,19**	**3,93**	**3,55**	**3,6**	**3,71**	**3,76**
Zn	mg/L	0,04	0,06	0,11	0,11	0,05	0,04	0,04	0,05
Cd	mg/L	0	0	0	0	0,01	0,01	0,01	0,02
Sr	mg/L	0,23	0,28	0,27	0,27	0,15	0,1	0,65	0,08
Ba	mg/L	1,13	0,94	1,96	1,96	0,88	0,57	4	0,27
Cu	mg/L	0	0	0,01	0,01	0,01	0	0,01	0

Ni	mg/L	**0,13**	0,08	**0,23**	**0,17**	**0,11**	**0,09**	**0,49**	0,1
As	mg/L	**6,12**	**6,24**	**6,1**	**6,33**	**6,13**	**6,39**	**7,02**	**7,19**
Co	mg/L	0,6	0,41	0,54	0,61	0,62	0,69	0,66	0,45
Pb	mg/L	**0,02**	**0,47**	**0,02**	**0,01**	0	0	**0,61**	0
K	mg/L	18,99	34,94	15,64	9,64	8,24	10,25	27,07	14,87
Mo	mg/L	0	0	0	0	0	0	0	0
Mn	mg/L	**1,27**	**1,68**	**4,39**	**2,01**	**0,69**	0,51	4,13	**0,89**
Cr	mg/L	0	0	0	0	0	0	0	0
Al	mg/L	0,04	**0,37**	**3**	**0,44**	**0,27**	0	0,17	0,02

- **Badalabougou:**

Analysis of the results in Table 17 reveals no contamination of the wells or boreholes sampled at Badalabougou. However, the pH of the wells and boreholes sampled at Badalabougou are acidic, with values ranging from 4.97 to 6.40, with an average of 5.69.

Table 18 shows the results of heavy metal analyses. The analysis reveals that the wells and boreholes are very rich in nickel, with values ranging from 0.03 to 0.06 mg/L, with an average of mg/L. Nickel exceeds the Malian standard of 0.02 mg/L. Borehole FB1 is rich in cadmium, with a value of 0.05, above the Malian standard. The latter exceeds the Malian standard of 0.02 mg/L. Borehole FB1 is rich in cadnium, with a value of 0.05, above the maximum permitted standard of 0.003 mg/L.

Table 17: Physico-chemical parameters of the Badalabougou well and borehole (commune V)

Parameter	Unit	FB1	PB1	PB2	FB2
pH	-	6,4	4,97	5,11	6,28
Conductivity	µS/cm	534	139	119	420
Turbidity	NTU	1	0	0	0
TA	-	0	0	0	0
TAC	-	220	140	100	260
Mg^{2+}	mg/L	25,92	5,76	8,64	43,2
Ca^{2+}	mg/L	40	11,2	28,8	44,8
Cl^-	mg/L	142	49,7	58,57	136,7
Fe^{2+}	mg/L	0,041	0,047	0,001	0
SO_4^{2-}	mg/L	4,31	0	0	1,21
NO_2^-	mg/L	0	0	0	0
NO_3^-	mg/L	8,04	25,79	25,61	6,17

Table 18: Heavy metal parameters for Badalabougou wells

Elements	Unit	FB1	FB2	PB1	PB2
Visit	mg/L	0	0	0	0
Zn	mg/L	0,11	0	0,68	0,27
Cd	mg/L	**0,05**	0	0	0
Sr	mg/L	0,05	0,04	0,02	0,02
Ba	mg/L	0,18	0,26	0,07	0,07
Cu	mg/L	0,03	0	0	0
Ni	mg/L	0,03	0,06	0,04	0,03
As	mg/L	0	0	0	0

Co	mg/L	0,72	0,37	0,21	0
Pb	mg/L	0	0	0	0
K	mg/L	3,68	9,76	3,89	5,55
Mo	mg/L	0,05	0	0	0
Mn	mg/L	0,28	1,02	0,03	0,04
Cr	mg/L	0,01	0	0	0
Al	mg/L	0	0	0	0

4.2. Discussions :

The results of the sorting of household waste in Bamako showed that wood constituted 17% of the categories of household waste most generated by the activities of the inhabitants of Bamako. This result is close to those found by (Bah, 2021) in his studies of household waste characterization in Faladjie. He states that among the components of household waste, vegetable waste accounts for 23.7%, which is one of the categories most generated in the activities of Bamako's inhabitants, just behind the mixture of sand, earth and ash, which accounts for 53.1%. Plastics come second with 16%, a result comparable to those of (Adamou, 2021). According to the latter the inhabitants of the city of Bamako use non-degradable plastic bags excessively despite the promulgation of Law No. 2014-024 of 03 July 2014 banning the production and marketing of non-biodegradable plastic bags. The plastic pollution entails serious health risks for wildlife in general and avian wildlife in particular (PRUBA, 2021).

The characteristics and mineralogical composition of the major elements and trace elements in the soil and in the sorted soil from the waste taken from the site are rich in iron oxide (Fe_2O_3), calcium oxide (CaO), potassium (K_2O), Cd, Mn, Cu, Zn. These values are higher than the maximum standard for ordinary soils and the French standards for soils taken from (Baize, 2000). As well as the levels found at crustal level by (Temgoua et *al*, 2015). These results are consistent with the findings of (PRUBA,2021), confirming that waste is rich in iron oxides, calcium oxide, aluminum and limestone. They are rich in zinc, copper and lead.

The soil profile is recognized as a filter that purifies groundwater, reduces the number of micro-organisms and retains a number of chemical elements (Aboubacrine, 1991). These contaminants can persist in the soil for varying lengths of time, depending on their nature. Simple ions, for example, persist in the medium for a long time, but can be precipitated in the form of salts and complexes by ion exchange, or interact with clays. (Biaou, 2019) confirms that high levels of heavy metals are probably due to the heterogeneity of the waste.

According to (Mehdi et al, 2007), landfill leachates are comparable to complex industrial discharges containing both organic and inorganic contaminants. Analysis of the leachate's physico-chemical parameters shows a slightly neutral to alkaline pH, ranging from 7.33 to 7.96 with an average of 7.59, in line with the Malian discharge limit value of between 6.5 and 9.5. These values are comparable to those of Hicham et *al*, (2021); Boucenna, (2007) and El Bada et *al*, (2010). Boucenna, (2007) showed that between November and January, the pH of the leachates varied from 7.2 to 7.9, which coincided exactly with our sampling dates. Conductivity values ranged from 16200 to 26000 μS/cm, with an average of 21950 μS/cm. The reference limit is between 14105 and 17100 μS/cm, which means that the leachate conductivity is very high, exceeding the WHO international standard and the Malian discharge standard of 2500 μS/cm. This conductivity value is close to that found by (Hicham et al, 2021). According to Malian legislation (discharge guide value 2500 μS/cm), under no circumstances should water be discharged into the environment. Leachate Turbidity values are all above one thousand (>1000) NTU and a very black coloration. This value corresponds to that found by (PRUBA, 2021). However, Mali has no specific threshold for the turbidity level of wastewater, but has adopted legislation that may be linked to it. $Ca^{(2+)}$ levels vary from 480 to 800 mg/L, with an average of 660 mg/L, while Mg^{2+} levels range from 576 to 1056 mg/L, with an average of 807.5 mg/L. Both values are well above the EU standard according to Moumouni et *al*, (2000). Chloride ion (Cl^-) values in the leachates range from 1597.5 to 6212.5 mg/L, with an average content of 4792.5 mg/L, well above the standard of 1200 mg/L recognized in Mali. Accidental discharge of chloride-rich leachate water into the environment can adversely affect the quality of soil, plants, animal life and fish, as well as the quality of traditional well water and surface water (gully water and water from the River Niger, despite its self-purifying properties) (PRUBA, 2021). Wage et *al*, (2000) assert that the chloride ion is a highly mobile element, easily migrating to the underlying water table. It is not affected by adsorption or ion exchange phenomena. Nor is it involved in acid-base or redox equilibria. Nor is it retained by soil clay-humus complexes (Chofqi, 2004). Leachate sulfate concentration is low, with a maximum value of 226.72 mg/L recorded in the Medina coura landfill leachate. This value is identical to that found by (Boucenna, 2007), who states that low sulfate levels are recorded in summer, when bacterial biodegradation phenomena are very intense (leachate heating) and environmental conditions are very reducing. Sulphates are then reduced to sulphides (H_2S), which are among the gases responsible for the unpleasant odours emitted by the landfill.

In Mali, the acceptable level of iron in wastewater is 2 mg/L or less. Iron levels obtained in leachate analyses fluctuate from 2.24 to 5.89 mg/L, with an average of 4.06 mg/L well above the Malian discharge standard. This explains the high concentration of iron in our leachate. The latter can infiltrate into the soil and pollute the water table. All trace metals are naturally present in trace amounts in the soil, and human activity may have reinforced this presence, according to (Djahida, 2016). The analysis results for heavy metals (Cu, Pb and Al) exceed the standard set in Mali. The copper concentration in samples taken at Dar-salam and Lafiabougou is measured at 0.15 mg/L, while at Badalabougou it is slightly higher at 0.35 mg/L. The acceptable discharge limit is set at 0.1 mg/L. Lead levels in leachates fluctuate between 0.31 and 1.58 mg/L. The acceptable discharge standard is set at 0.2 mg/L. Aluminum content ranges from 0.11 to 2.96 mg/L. The latter far exceeds the maximum value for wastewater discharges in Mali, which is less than or equal to 1 mg/L. These results are in line with the work carried out by (Djahida, 2016) for the case of the Tlemcen landfill and (Er-Raioui et al, 2011) for the case of Contamination of groundwater by leachate from public landfills: Case of the R'Mel water table (Province of Larache - North-West Morocco).

Looking at the physico-chemical analyses of the wells, we note that the pH of the wells (Medina-coura, Dar-salam, Lafiabougou, Badalabougou) varies from 4.23 to 6.75, with an average of 5.80. This is slightly below the Malian standard and the international WHO standard of between 6.5 and 9. This is slightly below the Malian standard and the international WHO standard of 6.5 to 9. This acidity of the water may be linked to the lithology of the aquifer, according to (Aboubacrine et *Al*, 1991). On the other hand, wells close to landfills, such as Pm1 Pm2 Pm3 in Medina-coura and PL2, PL3, PL4, PL5, PL6 in the Lafiabougou area, are much more acidic than distant wells Pm6 and PL8. According to (Mehdi et *al*, 2007), this difference may be due to the influence of CO_2 release from the landfill on groundwater. Conductivity values for the sampled wells ranged from 3.25 to 1488 µS/cm, with an average of 558.73 µS/cm. This average is below the maximum value of 1,500 µS/cm, which is the international standard set by WHO and Malienne. However, it should be noted that some wells, such as Pm1 and Pm5 in Médina-coura, as well as PL2, PL3, PL4 and PL5 in Lafiabougou, have a higher conductivity than the other wells and are close to the maximum standard. This indicates the presence of minerals and other potentially harmful ions in the water. Turbidity values range from 0.03 to 10.9 NTU, with an average of 1.80 NTU. Furthermore, the Médina-coura wells, in particular Pm1 and Pm6, show values close to the maximum permitted value, while Pm2 exceeded the Malian standard of 5 NTU. This explains why wells Pm1, Pm2 and Pm6 in Médine-coura, which are close to the landfill, show the presence of particles suspended in the water (organic debris, clays, micro-organisms etc.). Nitrate concentrations in the recorded wells range from 1.93 to 45.97 mg/L, with an average of 23.62 mg/L. The latter does not exceed either the Malian or the international WHO standard of 50 mg/L. On the other hand, the wells surrounding the Lafiabougou landfill sites, namely PL2, PL3, PL4 and PL5, which are very close to the landfill sites, stand out from the other wells for their high nitrate concentrations. Nitrate concentrations in these wells are very close to the Malian standard set (50 mg/L), with a maximum value of 45.97. This suggests that the wells are slightly contaminated with nitrate. These are generally the result of nitrification of organic nitrogen. This NO_3^- content indicates the presence of germs in the water (Mehdi et al, 2007). As for nitrite concentration, NO_2^- levels in wells surrounding landfills range from 0 to 1.43 mg/L, with an average of 0.59 mg/L. This average level is twice as high as the Mali and WHO standard of 0.2 mg/L. These values are in line with those found by (Boucena, 2007) in his study of the impact of the public landfill on groundwater quality at Zef-Zef (SKIKDA). Analysis of nitrite fluctuations in various wells shows that the Medina-coura wells: Pm2, Pm3, Pm4, PM5 and the Lafiabougou well: PL2 and PL5 exceeded the permitted standard. This explains the incomplete oxidation of ammonia (NH^{4+}) or reduction of nitrate (NO_3^-) in the water. The reduction of nitrate to nitrite can be attributed to bacterial activity. The presence of nitrite in the blood prevents hemoglobin from adequately binding oxygen, thus leading to the risk of acute methemoglobinemia. The presence of sulfates in drinking water can give it a perceptible taste and, at very high levels, sulfates can have a laxative effect in unaccustomed consumers. Taste alteration varies with the nature of the associated cation; taste detection thresholds range from 250 mg/l for sodium sulfate to 1000 mg/l for calcium sulfate. We generally consider taste alteration to be minimal at levels below 250 mg/l (WHO, 2017).

Analysis of heavy metals in the various wells shows a rise in Se, Ni, As, Pb and Al levels. This explains the presence of a metallic load in the well water. Heavy metals are toxic elements that can be harmful to human health. The wells are very rich in arsenic, with a maximum value of 7.19 mg/L. The maximum Malian standard is set at 0.01 mg/L. This explains the high contamination of wells with this element. Indeed, Arsenic presents an unacceptable risk index (greater than 1), for both adults and children (Moussa et al, 2018). However, it cannot be said from the results that waste played a role in arsenic pollution of the wells. Arsenic is a chemical substance naturally present in the environment. Its high concentration in the soil may be linked to its natural composition or to past or current industrial activities (HAS, 2020). According to Vincent et *al*, (2015) It is very difficult to define the cause or the parameters influencing arsenic levels. Observation of wells Pm1, Pm2, Pm3, Pm6 in medina-coura and PL2, PL3, PL4, PL5 in Lafiabougou shows contamination of the wells with Pb, Ni and Al. This result is in line with that found by Bouccenna, (2007) who states that groundwater is contaminated with metals mainly from point source pollution: landfill, urban and industrial areas. Lead and aluminum are one of the most toxic elements for humans (Tanouai et *al*, 2015). The presence of metals (Pb, Ni, Mn and Al) in well water (specifically the Lafiabougou wells) show contamination by waste.

V. CONCLUSION AND RECOMMENDATIONS

5.1. CONCLUSION :

Waste is one of the main sources of environmental pollution. Our study, based on the impact of waste on the quality of Bamako's groundwater, highlighted the major issues surrounding waste management in Bamako.

The city of Bamako has a household waste management problem. This mismanagement is harmful not only to the environment, but also to human beings. If left untreated, it can contain metallic charges that can contaminate the soil by burial and fermentation, and by percolation can penetrate and contaminate groundwater. Poor management of Bamako's household waste is increasing the quantity and type of waste generated. The different categories and fractions of household waste generate estimates that plastic waste is one of the most frequently generated wastes in the city of Bamako.

The physico-chemical characteristics of soils collected from landfills have demonstrated the importance of metal contaminants contained in waste in soil contamination. The high content of these metallic elements (cadmium, nickel, lead, zinc, copper and chromium) in the soil bears witness to the retention of metals contained in the waste.

The physicochemical characteristics of the leachate produced from the landfills show that the leachate contains high levels of chloride and nitrate, with higher turbidity and a high organic pollution load. The leachate has very high conductivity and chloride values. Consequently, the concentrations of dissolved ions and iron indicate that the leachate is anoxic and reduced water. The leachate therefore represents a potential source of contamination for the surrounding environment, and in particular for the water table at the landfill site.

Analyses of the heavy metal composition of the leachates revealed the presence of trace metals. Concentrations of Iron, Aluminium, Copper and Lead are abnormally high, due to the reception of waste containing these metals at the landfill. Contamination depends on several factors linked to the site's operating conditions (lack of leachate recovery and drainage systems, etc.).

Physico-chemical analyses of the wells revealed contamination of the wells located near the landfill, in particular the Medina-coura (Pml, Pm2, Pm3, Pm4) and Lafiabougou (PL2, PL3, PL4, PL5) wells. These tests revealed the presence of nitrite pollution and slight nitrate contamination. Assessment of well water quality by analysis also revealed the existence of certain pollutants found in the leachates examined. Bacteriological contamination was also detected, due to high concentrations of nitrites and nitrates, as well as the presence of heavy metals in the leachate and well water. Our physico-chemical analysis suggests that infiltration is the transfer vector for this pollution.

5.1.RECOMMENDATIONS :

As recommendations, we would like to :

- Development of transit depots in each district;
- Transport waste from transit depots to the final depot;
- Regular collection to avoid overflowing garbage at landfill sites;
- Monitor the environmental situation in the various communes of the Bamako district, in conjunction with the relevant structures;
- Raise awareness among people living near landfills not to consume untreated well water;
- Monitor the evolution of leachate and its impact on the underlying groundwater (choose wells close to the landfill at different depths from the piezometric level);
- Perform soil quality .

REFERENCES

Aboubacrine A, Amadou T, Adama M, Olivier B, Jean-Pierre V, Lahcen Ai, Mai. (1991). Hydrogeology and contamination of the water table supplying the city of Bamako (Mali).

Awa K. (2019). Waste management from the Marche de Banankabougou in commune VI of Bamako district.

Aouane M. and Bradai H. (2016). Bacteriological analysis of leachate from the corso technical landfill center before and after reverse osmosis (RO) treatment. Master's thesis, Université M'hamed Bougara de Boumerdès, p75.

Awa K. (2019). Waste management from the Marche de Banankabougou in commune VI of Bamako district.

BERTHE C. (2006). Study of the Organic Matter contained in leachates from different household and assimilated waste treatment plants. Doctoral thesis, University of Limoges.196 p.

Boucenna N. (2007). Impact des déchets de la décharge publique sur la qualité des eaux souterraines : cas de Zef-(Skikda), p57

Bouazza L. (2011). Effect of coagulation-flocculation on the quality of treated water from the Ain El Houtz WWTP. Magister thesis in chemistry, University of Tlemcen, 106 p.

C.A.E.P. (July 2018). Final report of the diagnostic study on the bottlenecks of large collectors and the routing of waste from transit depots to the final depot.

Chérifatou I B. (2019). Impact des Décharges Sauvages des Déchets Solides sur les Sols à Cotonou, Doi:10.19044/esj.2019.v15n30p94

Chunyan H. (December 12, 2016). Waste law: a comparative study of France and China.

CROUE J.P. (2001). Les matières organiques naturelles des eaux : Cours au Diplôme d'étude Approfondie " Chimie & Microbiologie de l'eau ". University of Poitiers.

Diabate M. (2010). Household waste: impact on health and environment in commune I of Bamako district: case of Banconi. Dissertation, Mali.

E. TEMGOUA, H. NTANGMO TSAFACK, H.-R. PFEIFER2 and T. NJINE3 (2015), Teneurs en élément majeurs et oligoéléments dans un sol et quelques cultures maraîchères de la ville de Dschang, Cameroun.

Elhafiane S. (2012). Solid waste management in the Urban Commune of Agadir and its impact on the natural environment.

Fettouch S. (2014). Membrane synthesis and applications in leachate treatment.

Gieguzynska E., Kocmit A. & Golebiewska D. (1998). Studies onhumic acids in eroded soils of Western Pomerania. In: Zaujec A., Bielek P.,Gonet S.S (Eds), Humic substances in Ecosystems. Slovak agricultural University, Nitra, 35-41.

HAS, (July 2020). Arsenic-polluted soils: Information for residents.

H. Er-Raioui, S. Bouzid, S. Khiannous and M. A. Zouag (June 2011). Groundwater contamination by landfill leachate: Case of the R'Mel water table (Larache Province - North-West Morocco).

Hakmi A. (2006). Traitement des eaux " traitement de l'eau de source bousfer Oran. Université des sciences et de la technologi p 76.

Herley R J. (2005). Typology and analysis of municipal waste management (household and market waste) in the city of Cayes, Haiti.

Hicham K, Lotfi A, Christiane L, Jacky M. (2001). Évaluation de l'impact des lixiviats d'une décharge d'ordures ménagères sur la qualité physico-chimique et bactériologique des eaux d'un ruisseau de Franche-Comté. Environnement, Ingénierie & Développement, N°24 - 4th Quarter 2001, pp.3-6. http://www.amblans-et-velotte.fr/index.php?IdPage=1349182733

Issa O, Ahamadou D, Yakouréoun D, Oumar D and Adama K. (2017). Actors and strategies for household solid waste management in Bamako.

Jérémy V. (2015). Study of arsenic contamination in borehole water: Case of Seno province, Burkina Faso. Journal officiel de la république du Mali. LOI N°2021-032 DU 24 MAI 2021 Relative aux pollutions et aux nuisances.

Labadi N. and Saadoudi M. (2019). Contribution à l'étude de caractérisation des lixiviats du center d'enfouissement technique d'El Oued (CET). Master's thesis, Biodiversity and environment, Université Echahid Hamma Lakhdar D'el-Oued, p88.

Mamadou K. (2018). Urban typology and potential geographical accessibility of so-called "modern" health facilities in the district of Bamako (Mali).

Mehdi. Mekaikia Mokhtarial, Belabbed Bourhan Eddine2 Djabri. Larbi, Hani Azzedine Laour Rabah (2007). Characteristics of the Tiaret city landfill and its impact on groundwater quality. p93, 99.

Djahida M. (2016). The risks of pollution of the natural environment by leachate from controlled landfills. Cas du center d'enfouissement technique de Tlemcen. P18

ADAMOU M. (June 12, 2021). Caractérisation des ordures ménagères et risques Microbiologies potentiels associes aux décharges située à proximité des Habitations à Bamako p12, p20, p51, p151

Moumouni W, Abdoul R, Zanguina A. (2000). Caractérisation Physico-Chimique Des Lixiviats Des Décharges: Cas De La Décharge Non Contrôlée De Niamey 2000 (Niamey-Niger).

Moussa B, Mounira T, Marwa A, Abdelaziz L. (2018). Groundwater contamination by metallic trace elements in the azzaba region of algeria.

WHO (World Health Organization). (2017). Guidelines for drinking-water quality: 4th ed. incorporating first addendum [Guidelines for drinking-water quality: 4th ed. incorporating first addendum].

Oumar B. (2021). Caractérisation des déchets solidesménagers de Faladié, p10.

Ouriba S. (2016**).** Roles and perspection of stakeholders in sustainable solid waste management in developing countries: the case of Commune IV of the District of Bamako.

Quesnel P. (July 2001). Artificial wetland treatment of leachate from a trench deposit in a northern environment.

PRUBA. (April 2021). Environmental and social audit of the Noumoubougou Final solid waste landfill in Mali p45.

Seydou D, Dogodiougo D, Fousseyni D. (February 2018). Etat des lieux de la gestion des déchets District de Bamako. Rapport Final.

Sofiane A, Zerdi F. (2022). Impact of solid waste landfill on groundwater quality in the Ain del Fakroun region (Oum El Bouaghi).

Tanouayi G, Gnandi K, Ahoudi H, Ouro S K. (March 2015). Metallic contamination of surface and groundwater in the Hahotoe-Kpogame phosphate mining area (South Togo): the case of cadnium, lead, copper, and nickel.P35-50.

Yacouba M. (2018). Electronic and computer equipment waste in Mali: Quantification and recovery. PhD thesis, USTTB, p 252

ZEDAM A S, LARABA Z0 . (2019/2020). Comparative study of leachate characteristics: source of pollution and contamination, in four landfills p4, p7, p9, p16.

Appendices

Appendix 1: STANDARDS :

Tab: French soil standards

Ref ; Denis Baize: Total heavy metal content in French soils. Courrier de L'environnement de l'INRA nO39 February 2000

Metals	Cu	Zn	Pb	Ni	Li	Cr	Co	Ca	Mg	Na	K	Mn	Cd	Fe
Sampling location	ppm	ppm	Ppm	ppm	ppm	ppm	ppm	%	%	%	ppm	ppm	ppm	%
Standards for ordinary floors	30	50	10	40	20	100	8	1.37	0.50	0.63	0.83	600	-	0.38
Floor standards	100	300	50	50	30	150	-	-	-	-	-	30	2	-

Malian water standards established by the Malian standards agency AMANORM

DRINKING WATER STANDARDS

Physico-chemical characteristics

Parameters	Units	Max. permissible values
Color	Platino-cobalt or Hazen	25
Turbidity	NTU	5
$KMnO_4$ oxidizability (hot)	mg / L	5
pH	-	$6.5 \leq pH \leq 9$
Chlorides	mg / L	250

Sulfates	mg / L	500
Total hardness	mg /L (C_aCO_3)	500
Calcium	mg / L (C_aCO_3)	200
Magnesium	mg / L (C_aCO_3)	100
Sodium	mg / L	200
Potassium	mg /L	100
Total dissolved solids	mg / L	1200
Electrical conductivity	µS/cm	1500
Nitrates	mg/ L of (NO_3^-)	50
Nitrites	mg/ L of (NO_2^-)	0.2
Ammonium	mg/ L of (NH_4)	0,5
Aluminum	mg / L	0,2
Iron $(Fe)^{++}$	mg /L	0.3
Manganese	mg /L	0,5
Copper	mg /L	1
Zinc	mg /L	1
Fluorides	mg /L	1 ,5
Molybdenum	mg /L	0,07

Heavy metals

Parameters	Units	Maximum permissible values
Arsenic	mg /L	0,01
Cadmium	mg / L	0,003
Total chromium	mg / L	0,05
Cyanide	mg / L	0,07

Mercure	mg / L	0,001
Lead	mg / L	0,01
Selenium	mg / L	0,01
Nickel	mg / L	0,02
Antimony	mg / L	0,005
Barium	mg / L	0,7
Boron	mg / L	0,5
Aluminum	mg / L	0.2

Malian wastewater discharge standards :

N°	**Parameters**	Permissible limit values
01	**PH**	6.5 - 9.5
02	**Temperature (°C)**	≤ 40
03	**Color**	The spill must not cause
		no color of the receiving environment
04	**Odor**	The spill must not cause modification of the receiving environment
05	**Floating matter (> 1 cm)**	Absence
06	**Settling solids (mg/L)**	<0.5 after 2 hours of settling
07	**Suspended solids (mg/L)**	≤ 30
08	**Biochemical oxygen demand for 5 days BOD5 (mgO2/L)**	≤ 50
09	**Chemical oxygen demand COD (mgO2/L)**	≤ 150
10	**Dissolved oxygen**	> 6mg O_2/L
11	**Aluminum (mg/L)**	≤ 1
12	**Arsenic (mg/L)**	≤ 0,5

13	**Cadmium (mg/L)**	≤ 0,02
14	**Chromium III (mg/L)**	≤ 2
15	**Chromium VI (mg/L)**	≤ 0,2
16	**Iron (mg/L)**	≤ 2
17	**Manganese (mg/L)**	≤ 2
18	**Mercury (mg/L)**	≤ 0,005
19	**Nickel (mg/L)**	≤ 2
20	**Lead (mg/L)**	≤ 0,2
21	**Copper (mg/L)**	≤ 0,1
22	**Tin (mg/L)**	≤ 10
23	**Zinc (mg/L)**	≤ 0,5
24	**Cyanide (mg/L)**	≤ 0,5
25	**Chlorine (mg/L)**	≤ 0,2
26	**Sulfides (mg/L)**	≤ 1
27	**Sulfites (mg/L)**	≤ 1
28	**Sulfates (mg/L)**	≤ 1000
29	**Chlorides (mg/L)**	≤ 1200
30	**Fluorides (mg/L)**	≤ 6
31	**Phosphorus (mg/L)**	≤ 10
32	**Ammoniacal nitrogen (NH4mg/L)**	≤ 15
33	**Nitrates (mg/L)**	≤ 30
34	**Nitrite (mg/L)**	≤ 0,6
35	**Oils and fats (animal + vegetable) (mg/L)**	≤ 20
36	**Mineral oils (mg/L)**	≤ 5
37	**Phenols (mg/L)**	≤ 0,5
38	**Aromatic solvents (mg/L)**	≤ 0,2
39	**Nitrogenous solvents (mg/L)**	≤ 0,1
40	**Chlorinated solvents (mg/L)**	≤ 1

41	**Surfactants (mg/L)**	≤ 2

Appendix 2: ANALYSIS PROTOCOLS

PHYSICAL PARAMETERS

1- pH: hydrogen potential :

The pH (hydrogen potential) is one of the fundamental characteristics of water. pH is a measure of acidity, i.e. the concentration of hydrogen ions (H^+), or basicity, i.e. the OH^- ions contained in a substance.

In practice, the pH scale ranges from 0 (very acidic) to 14 (very alkaline): the median value of 7 corresponds to a neutral solution at 25°c. The pH of natural water can vary from 4 to 10, depending on the acidic or basic nature of the soil.

The pH value must be taken into consideration in most water treatment operations, especially when a chemical reaction is involved, and because some processes require a specific pH value to be effective. pH is also I one of the most important parameters for determining water quality. It is used to control the quality of water entering the treatment plant (significant variations in pH are almost always the result of industrial discharges).

In particular, low pH levels (acidic water) increase the risk of metals being present in a more toxic ionic form.

High pH levels increase ammonia concentrations, which are toxic to fish. This parameter characterizes a large number of physico-chemical equilibria and depends on multiple factors, including the origin of the water.

1.1 Principle :

The method is based on the use of a pH meter. The pH meter is a rather unusual voltmeter characterized by a very high input impedance due to the high resistance presented by I measuring electrode.

1.2 Apparatus :

The pH measurement equipment consists of :

- A HANNA HI 2209 pH meter equipped with a combination electrode:
- Built-in thermometer

- A magnetic stirrer.

1.3 Reagents :

-Buffer pH = 4.01

- Buffer pH = 7.00

1.4 Procedure :

a. Preparation of I instrumentation

- Check all connections: mains, electrodes, etc;
- Remove the protective cap from the double electrode and store in a safe place
- Rinse electrode tip thoroughly with distilled water, then wipe dry with toilet paper.

b. Calibration

- Switch on pH meter,
- Introduce a small magnetic bar into a perfectly clean measuring vessel (a small beaker, for example);
- Rinse vase and rod with clean distilled water;
- Add pH 4.01 buffer to the vessel, then immerse the electrode, ensuring that the magnetic rod can rotate freely
- Wait until the reading is stable;
- Rinse the measuring vessel and electrode with distilled water, then introduce the second buffer (pH 7.00) into the vessel in the same way as before, immerse the electrode and wait for the reading to stabilize;
- The device is now calibrated and ready for use.

c. pH measurement

- Rinse the vessel, magnetic bar and electrode with distilled water, then with the sample;
- Fill the measuring vessel with I sample;
- Immerse the electrode with the usual precautions and gentle agitation:
- Read pH directly when value has stabilized.

2. CONDUCTIVITY:

The conductivity of water is a measure of its ability to carry electric current. Most of the substances dissolved in water are in the form of electrically-charged ions, so measuring conductivity enables us to assess the quantity of salts dissolved in the water. It is therefore logical to assume that the higher the mineral content of the water, the greater its conductivity.

There is a relationship between the conductivity of water and its mineralization, hence the importance of measuring conductivity, an almost instantaneous measurement, to determine the mineralization of water.

2.1. Principle of measurement :

Conductivity is measured using a conductivity meter.

2.2 Procedure :

a. Calibration:

Thoroughly rinse the conductivity meter probe with distilled water and wipe dry with toilet paper. Immerse the probe in the 1413 µS/cm conductivity standard and wait for the displayed value to stabilize perfectly. As soon as the conductivity meter has stabilized with this standard, rinse the probe and proceed to use the instrument on the samples to be analyzed.

b. Conductivity measurement:

Whatever the device used :

- Check cell/conductimeter connections. Rinse a container and a magnetic rod with distilled water, then with the sample;
- Rinse the cell thoroughly with distilled water and wipe dry with toilet paper;
- Place a sufficient quantity of the sample to be analyzed in the container containing the magnetic rod, then immerse the conductivity meter's electrode (or cell).
- Use gentle agitation to ensure that the ionic concentration between the electrodes is identical to that of the surrounding liquid. This agitation also eliminates air bubbles on the electrodes.

- Read conductivity directly when the value has stabilized. It is expressed in µS/cm.

NB: When using the HI 8633 multifunction instrument (pH, conductivity, temperature and salinity), press M to select pH or conductivity mode.

Figure 30: pH and conductivity measurements of leachate and well water

3. TURBIDITY :

Turbidity reduces the transparency of a liquid. It therefore reflects the presence of particles suspended in the water (organic debris, clay, micro-organisms, etc.).

Turbidity plays a very important role in water treatment. This is because :

- it indicates a higher probability of the presence of pathogenic elements;
- turbidity interferes with disinfection, ultraviolet treatment is ineffective and chlorine treatment also loses its effectiveness;
- the organic matter associated with turbidity favors the formation of biofilms in the network, and consequently the development of bacteria that are insensitive to chlorine in particular.

5.2.Principle :

The method involves determining the turbidity ratio using a hephelometric light scattering signal versus a transmitted light scattering signal, based on a standard range of formazine solutions.

5.3.Reagents :

o Standard solutions 10, 20, 100 and 800 NTU

3.3 Apparatus:

- Turbidimeter HACH 2100Q is:
- Transparency evaluation cuvette made of 10 mL colorless glass.

3.4 Procedure :

a. Calibration: with StablCal standards

- Press the calibration button to enter calibration mode and follow the instructions on the display:

NB: carefully find each standard before inserting it, the standard to be inserted is framed.

- So, following these instructions, insert the standard (StablCal 20 NTU) into the cuvette holder, close the lid and press Measure The display shows Stabilization followed by the result;
- Insert the StablCal 100 NTU and 800 NTU standards in the same way, following the instructions on the display.
- Press Done to complete calibration.
- On completion of calibration, the instrument automatically switches to Verify calibration mode, i.e. verification of calibration with the 10 NTU standard. Insert standard and measure to verify calibration.
- After this, the device is ready for the recommended analysis.

b. Turbidity measurement :

- Fill the measuring tube with the homogenized sample, then wipe clean with toilet paper;

- Insert the measuring tube into the chamber (cuvette holder), close and press Measure ;
- Read the result directly after stabilization.

3.5. Expression of results:

Sample turbidity is the value given (in NTU) by the instrument.

4. ALKALINITY (TA) AND TOTAL ALKALINITY (TAC)

Water alkalinity corresponds to the presence of bicarbonate, carbonate and hydroxide ions.

4.1. Principle :

This is the determination of successive volumes of strong acid in dilute solution, required to neutralize the volume of water to be analyzed at pH levels = 8.3 and 4.5. The first determination is used to calculate the alkalimetric titre (TA), the second to calculate the complete alkalimetric titre (TAC). As a result :

- water with a pH below 8.3 contains no TA;
- water with a pH below 4.5 contains neither TA nor TAC.

4.2 Reagents :

- HCI 0.1 N
- phenolphthalein 1% (m/v ethanol)
- helianthin

4.3. Equipment :

o A burette

o A 50 mL test tube

o A dropper

4.4 Procedure :

TA

- Place 50 mL of water to be analyzed in a beaker;
- Add 4 drops of phenolphthalein;
- If the color changes to pink, titrate with hydrochloric acid (HCI 0.1N) until colorless and note the volume.
- But if there was no phenolphthalein turn, this means that the TA is zero.

Expression of results

$$TA = \frac{VaxNax1000}{V} x100$$

TAC:

Add 4 drops of helianthine to the same test sample used for TA, and titrate with hydrochloric acid until the color changes from yellow to red. Note the volume poured

Expression of results :

$$TAC = \frac{VaxNax1000}{V} x100$$

TA and **TAC** are expressed in mg/L

Va: volume of HCl used for dosing

Na: normality of HCI (0.1 N)

V: the test run

1000 is the conversion factor from liter to

100, is the conversion factor from mmol/L to mg/L

5. TOTAL/CALCIUM HARDNESS:

Water hardness is the sum of metal cation concentrations, excluding alkali metals (Na+, K+) and H+. In most cases, it is mainly due to Ca++ and Mg++ ions. The presence of these two cations in water often reduces the toxicity of metals, but their excess prevents the formation of foam with soap. Hardness is measured in mg CaCO3 per liter.

5.1. Principle:

The principle is a complexometric molar titration with I EDTA, of Ca++ and Mg++ ions at pH10. The colored indicator, black Eriochrome T (for Mg++) or patton reager (for Ca++), which gives a dark red color, is used as a light-blue conversion indicator.

5.2. Reagents :

- EDTA..0.005 mol/L
- Sodium hydroxide solution...............................2 mol/L
- Ammonia buffer pH 10
  - ammonium chloride NH4CI..................................5,4g
  - pure ammonia...44 mL
  - final volume (with distilled water)...............................100 mL

5.3. Equipment :

- a burette
- 25-50 ml test tube or 25 ml volumetric pipette
- A 10 ml graduated pipette

5.4. Procedure :

a. total hardness: (Mg2+, Ca2+)

- take 25 mL of the sample to be analyzed, - take the sample to be analyzed, - take the sample to be analyzed, - take the sample to be analyzed.
- add 2.5 mL of ammonia buffer solution pH10,
- then a spatula tip of black Euriochrome T.
- titrate with EDTA until red turns blue.

Hardness (total) = VN 24 25 1000

b. calcium hardness: (Ca2+)

- take 25 mL of I sample to be analyzed,
- add 5 mL NaOH solution,
- then a tip of Patton Reeder's spatula,
- EDTA until the colour changes from red to blue.

5.4.**Expression of results** :

Hardness (total): $= \frac{V*N*100}{PE} \text{x}1000$

VD: Volume required for dosing,

N: normality of | EDTA (0.01 N)

PE: Test sample 100 is the conversion factor from mmol/L to mg/L $CaCO_3$,

1000 is the conversion factor from liter to mL.

VI. CHLORIDE ION DOSAGE: $Cl^{(-}$

The chloride content of water depends on the origin of the water and the terrain through which it flows.

6.1. Principle:

The reaction of chloride ions with silver ions ($AgNO_3$) forms insoluble silver chloride, which is precipitated quantitatively. Potassium chromate (or dichromate) is used as a red-tone indicator (Ag_2CrO_4).

$AgNO_3 + NaCl^- \rightarrow AgCl + NaNO_3$

$2AgCl + K_2CrO_4 - 2KCl + Ag_2CrO_4$

6.2 Reagents :

- Silver nitrate solution 0.025 mol/L :
  - Silver nitrate-...4.247 g
  - Final volume (with distilled water).......................000mL
- Potassium chromate or dichromate 10% (m/v distilled water)

6.3. Equipment:

- A burette
- A 50 mL test tube
- A dropper

6.4. Procedure :

To determine chloride ions, take 50 mL of sample and place in a beaker or Erlenmeyer flask. Add 2-3 drops of potassium chromate (or dichromate), and the solution turns yellow. Titrate with silver nitrate solution until yellow turns red.

6.5. Expression of results :

$$Cl = = \frac{Vb * CAgNO3 * MCL-}{PE} \text{x}1000$$

Cl: is expressed in mg/L (1000, indicates conversion from g to mg)

Vb: volume of $AgNO_3$ used for assay:

CAgNO3: $AgNO_3$ concentration (0.025 M):

Mc-: molar mass of chlorine (35.5 g/mol);

PE: test sample (50mL):

7. DETERMINATION OF SULFATE IONS: SO_4^{2-}

7.1. Principle:

The precipitation of sulfate ions in the barium sulfate state in the presence of barium chloride determines the sulfate ions present in the water.

$BaCl_2 + SO_4^{2-}$ $BaSO_4 + 2Cl^-$

7.2. Reagents :

- 10% hydrochloric acid solution :
- HCI concentrate 37%..27mL
- Final volume (with distilled water)........................100mL
- Stabilized barium chloride solution:
- barium chloride..10g
- TWeen solution (PVP 20%)........................... 20Ml
- final volume (with distilled water)........................100mL

7.3. Equipment:

UV Visible spectrophotometer.

7.4. Procedure

- Place 7.8mL of sample in a test tube;

- Add 7.8 mL distilled water to another tube for the control; add 0.2 mL hydrochloric acid solution and 1 mL stabilized barium chloride solution to each tube;

- Shake and let stand for at least 15 minutes to take spectrophotometer readings at 650 nm wavelength. The reference is distilled water.

7.5. Expression of results

The relationship used to calculate the sulfate ion content is given by :

$$S0_4^{2-} = = \frac{ABS-0{,}0119}{0{,}0116}$$

ABS is the absorbance read by the spectrophotometer

UNDESIRABLE SUBSTANCES

1- DETERMINATION OF NITRITE IONS: NO2-

Nitrites (NO2) are undesirable compounds in the composition of drinking water. Its presence may be due to the incomplete oxidation of NH4 or to the reduction of NO3 present in the water. The reduction of NO3 to NO2 may be due to bacterial action. Water containing NO ions may be suspected of having recently been contaminated by faecal matter.

The presence of nitrites in the blood prevents hemoglobin from adequately binding oxygen, thus leading to the risk of acute methemoglobinemia. Nitrites are also highly toxic to fish, and often fatal. For this reason, the nitrite content of drinking water is regulated, and indirectly that of nitrates, due to their ability to transform into nitrite. Guide and maximum permissible nitrite concentrations in water intended for human consumption are 0.1 mg/l and 1 mg/l respectively. A concentration of 1 mg/l is a sign of pollution. A microbiological analysis should then be carried out. Given the reuse of wastewater in agriculture and/or fish farming, it is also important to determine the nitrite content of wastewater.

1.1. The principle:

Nitrites are determined spectrophotometrically at 435 nm, using the ZAMBELI method. Sulfanilic acid in hydrochloric acid medium, in the presence of ammonium ions and phenol (= ZAMBELLI reagent), forms a yellow complex with nitrite ions, the intensity of which is proportional to the nitrite concentration.

1.2 Reagents :

- Pure ammonia (concentrate);
- Zambelli reagent :

- Concentrated HCI................................260mL
- Sulfanilic acid...........................5g
- Crystallized phenol............................-7.5g
- Ammonium chloride......................135g
- Distilled water (NO2-free)-..........625mL
- Final volume (with distilled water) -.......1000mL

1.3 Apparatus:

UV Visible spectrophotometer.

1.4 Procedure

Place 10 mL of sample to be analyzed in a tube;

Add 10 mL of distilled water to another tube for the control.

Add 0.4 mL Zambelli's reagent to each tube; shake and wait about 10 minutes to add 0.4 mL pure ammonia;

Shake and let the reaction develop for 15 minutes, then take a spectrophotometer reading at 435nm wavelength. The reference is distilled water.

1.5. Expression of results

$$NO2- = \frac{ABS}{0,0045}$$

Expressed in mg/L

ABS is absorbance read by spectrophotometer

Preparation of Zambelli reagent :

- Place distilled water and hydrochloric acid in a 1-liter volumetric flask:
- Dissolve the sulfanilic acid and phenol in the mixture, heating gently in a bain-marie until completely dissolved:
- Add ammonium chloride and stir until dissolved;
- After cooling, make up to 1 liter with distilled water.

Figure 31: Color change indicating the presence of nitrite

2- DETERMINATION OF NITRATE IONS: NO3-

$NO^{(3-)}$ ions are generally the result of nitrification of organic nitrogen, or come from the dissolution of fertilizers. Nitrate determination involves relatively complex methods, with a high probability of the presence of interfering constituents, making nitrate determination a delicate process.

The proposed spectrophotometric method requires a clear sample: the sample must be filtered.

2.1 Principle:

In the presence of sodium salicylate, nitrates give sodium paranitrosalicylate, colored yellow and suitable for spectrophotometric determination.

Figure 32: Color change indicating the presence of nitrate

2.2 Reagents :

- 0.5% sodium salicylate solution:
- 1 mol/L sodium hydroxide solution;
- Sodium potassium tartrate solution:

- Sodium hydroxide ------400g
- Double sodium and potassium tartrate......60g
- Final volume (with distilled water) -...............100ml

2.3 Apparatus:

UV Visible spectrophotometer.

2.4 Procedure

The nitrate ion content is calculated as follows:

- Dispense 10mL of sample into a flat-bottomed flask;
- At the same time, take 10 mL of distilled water for the control; add 1 mL of sodium salicylate solution plus a drop of NaOH:
- heat until evaporated to dryness, then allow to cool sufficiently; dissolve the dry residue with 2mL of concentrated sulfuric acid and wait about 10 minutes: finally add 15mL of distilled water and 15mL of the sodium-potassium double tartrate solution,

- allow reaction to develop for approx. 15 min., then proceed to spectrophotometer reading at 415 nm wavelength. The reference is distilled water.

2.5 Expressing results

The following relationship gives the nitrate ion content:

$$\frac{ABS}{0{,}0497}$$

ABS is absorbance read by spectrophotometer

3 DETERMINATION OF IRON (Fe^{2+}): orthophenanthroline method

3.1 Principle:

The addition of phenanthroline 1.10 solution to a test sample, after reduction of the Fe^{3+} to Fe^{2+} forms by hydroxylamine hydrochloride, gives a red-orange complex in a medium buffered by ammonium acetate.

3.2 Reagents:

All reagents must be prepared from distilled water, free from any trace of iron.

- 10% hydroxylamine chloride solution (w/v distilled water) :
- Phenanthroline solution 0.42% :

- phenanthroline 1.10........................-0.42g
- concentrated hydrochloric acid.............2 drops
- Final volume (with distilled water) 100m

Acetate buffer solution:

- ammonium acetate...........................40g
- acetic acid................................. 50mL
- Final volume (with distilled water).............100mL

3.3 Apparatus:

UV Visible spectrophotometer.

3.4 Procedure:

This determination involves adding 50 mL of the sample to be analyzed:

- 1mL of hydroxylamine chloride solution:

- 2mL acetate buffer solution:
- 2mL of phenanthroline solution;
- Keep the mixture in the dark for 15 min, then take a spectrophotometer reading at 510 min wavelength. The reference is distilled water.
- At the same time, make a control by replacing the sample with 50 mL of distilled water.

3.5 Expressing results

The relationship used to calculate the ferrous ion content is given by

$$Fe2+=\frac{ABS-0,0256}{0,3009}$$

Expressed in mg/L

ABS is the absorbance read by the spectrophotometer.

Figure 33: Image of leachate sampling at the Lafiabougou landfill (Commune IV)

Ousmane DOUCOURE

IMACT OF SOLID WASTE ON WATER QUALITY

SOUTERRAINES DU DISTRICT DE BAMAKO

Summary

In Bamako, the production of household waste is intensifying with population growth, and can have strong impacts on the environment, such as pollutants like leachates, which can infiltrate and pollute groundwater and watercourses. In this study, based on the impact of waste on groundwater quality in the Bamako district, several methods and tools were used: an inventory of the various transit dumps in the six communes of the Bamako district, waste characterization, physico-chemical and heavy metal analyses of soil, landfill leachates and well water surrounding the said dumps. The results of these analyses showed that the waste is composed of several elements. The physico-chemical analysis of the soil, leachates and wells surrounding the landfills showed that these elements do not comply with Mali's and WHO's international standards. The proposed strategies could help to raise awareness among people living near landfills, so that they do not consume untreated well water.

Keywords: Waste, Leachate, pollution, impact, Bamako

Abstract

In Bamako, the production of household waste intensifies with population growth and can have strong impacts on the environment such as: pollutants such as leachate which can infiltrate and pollute groundwater and waterways. Thus, within the framework of this study based on the impact of waste on the quality of groundwater in the district of Bamako, several methods and tools were used: the inventory of the various transit deposits in the six communes of the district of Bamako, characterization of waste, physico-chemical and heavy metal analyses of soils, leachate from landfills and water from wells surrounding said landfills. The results obtained from these analyzes showed that the waste is composed of several elements. The physico-chemical analysis of the soils, leachates and wells surrounding the said landfills showed that these elements do not meet the standards accepted in Mali and the international OMS. The proposed strategies could contribute to Sensitize populations living in the vicinity of landfills not to consume water from wells without treatment.

Key-words: Waste, leachate, pollution, impact, Bamako

MIX
Papier aus verantwortungsvollen Quellen
Paper from responsible sources
FSC® C105338

Printed by Books on Demand GmbH, Norderstedt / Germany